The
NATURAL HIST
of
PENNSYLVANIA

Written by Stan Freeman
Illustrated by Mike Nasuti

Pennsylvania State Capitol in Harrisburg

Barn owl *Monarch butterfly* *Bobcat* *Daisy fleabane*

Hampshire House Publishing Co.
Florence, Mass.

Red fox

CONTENTS

American toad

Great blue heron

Garter snake

Eastern chipmunk

Red-tailed hawk

Raccoon

Water lily

Pennsylvania
Quick facts

Capital: Harrisburg

Origin of name: In honor of William Penn

Nickname: Keystone state, Quaker state

Population (2014 est.): 12,787,209
  (Sixth among all states)

Total area: 46,055 square miles
  (33rd among all states)

Geographic center: In State College in Centre County, 2.5 miles southwest of Bellefonte

State forest preserves: About 2.2 million acres

Miles of rivers and streams: 83,260 miles

Inland lakes and ponds: About 4,000

Highest mountain: Mount Davis, 3,213 feet above sea level, located in Forbes State Forest in Somerset County

Highest waterfall: Raymondskill Falls, located in Bushkill, with three tiers that have a combined height of about 150 feet

Largest natural lake: Harveys Lake in Luzerne County is the largest natural lake by volume, capable of holding 7.7 billion gallons. The largest by surface area is Conneaut Lake in Crawford county, about 930 acres.

Longest river: Ohio River, 981 miles. It starts in Pittsburgh where the Allegheny and Monongahela rivers join. It then flows through six states to Cairo, Illinois, where it empties into the Mississippi River.

State symbols

Animal: white-tailed deer

Fish: brook trout

Game bird: ruffed grouse

Tree: eastern hemlock

Flower: mountain laurel

Insect: Pennsylvania firefly

A brief history of Pennsylvania

The first humans in Pennsylvania were likely hunters and gatherers who entered the state as the last ice age was ending, perhaps 13,000 years ago to as early as 19,000 years ago. They came to hunt woolly mammoths and other big game.

Native American tribes eventually became established throughout Pennsylvania, including the Lenape (also called the Delaware), Shawnee, Iroquois, Susquehannock and Erie.

The first European to explore the region may have been Giovanni da Verrazzano during his 1524 voyage for the French. A map from that expedition survives on which the Pennsylvania region is referred to L'arcadia, or "wooded coast." In 1609, the English explorer Henry Hudson, employed by the Dutch East India Company, sailed into Delaware Bay, and in 1616, the Dutch explorer Cornelius Hendrickson sailed up the bay to the mouth of the Schuylkill River.

One of the most important documents in history, the Declaration of Independence, was adopted in Philadelphia on July 4, 1776, during a gathering of American colonial representatives – the Continental Congress. Pennsylvania ratified its state constitution on September 28, 1776, and was granted statehood on Dec. 12, 1787, the second state to sign the U.S. Constitution, five days after Delaware.

Pennsylvania's population

Year	Population	Year	Population
1800	602,365	1920	8,720,017
1820	1,049,458	1940	9,900,180
1840	1,724,033	1960	11,319,366
1860	2,906,215	1980	11,863,895
1880	4,282,891	2000	12,281,054
1900	6,302,115	2014 est.	12,787,209

Pennsylvania's largest cities (as of 2014)

City	Population	City	Population
Philadelphia	1,560,297	Bethlehem	75,135
Pittsburgh	305,412	Lancaster	59,302
Allentown	119,104	Harrisburg	49,082
Erie	99,452	Altoona	45,558
Reading	87,812	York	43,865
Scranton	75,281	State College	42,100

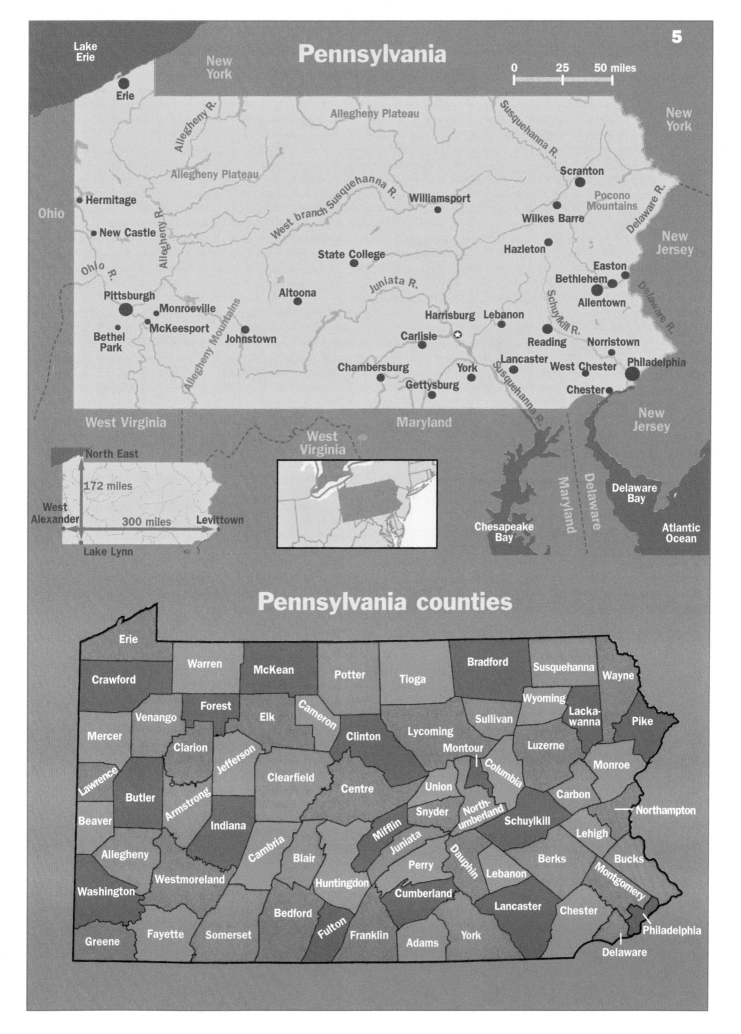

Pennsylvania

Lake Erie

New York

0 25 50 miles

New York

Erie

Allegheny Plateau

Susquehanna R.

Scranton

Pocono Mountains

Ohio

Hermitage

Allegheny R.

West branch Susquehanna R.

Williamsport

Wilkes Barre

New Jersey

Delaware R.

New Castle

State College

Hazleton

Easton

Bethlehem

Ohio R.

Juniata R.

Altoona

Harrisburg ✪ Lebanon

Allentown

Schuylkill R.

Pittsburgh

Monroeville

Allegheny Mountains

Reading

Norristown

Delaware R.

Bethel Park

McKeesport

Johnstown

Carlisle

Lancaster

West Chester

Philadelphia

Chambersburg

York

Gettysburg

Susquehanna R.

Chester

West Virginia

Maryland

New Jersey

North East

West Virginia

172 miles

West Alexander

300 miles

Levittown

Lake Lynn

Delaware Bay

Chesapeake Bay

Maryland

Delaware

Atlantic Ocean

Pennsylvania counties

Erie

Warren

McKean

Potter

Tioga

Bradford

Susquehanna

Wayne

Crawford

Wyoming

Lacka-wanna

Pike

Venango

Forest

Elk

Cameron

Sullivan

Mercer

Clarion

Clinton

Lycoming

Montour

Columbia

Luzerne

Monroe

Lawrence

Jefferson

Clearfield

Centre

Union

North-umberland

Carbon

Northampton

Butler

Armstrong

Indiana

Snyder

Schuylkill

Beaver

Mifflin

Juniata

Lehigh

Allegheny

Cambria

Blair

Perry

Dauphin

Berks

Bucks

Washington

Westmoreland

Huntingdon

Cumberland

Lebanon

Montgomery

Greene

Fayette

Somerset

Fulton

Bedford

Franklin

Adams

York

Lancaster

Chester

Philadelphia

Delaware

A little more than a century ago, America was not a land where the deer and the antelope played or the buffalo roamed. Populations of these animals and others had been reduced to the point that the survival of some species was very much in doubt.

Similarly, parts of the Northeast were nearly barren of many species of wildlife that had lived here for thousands of years, including black bears, moose and beavers.

The cutting of forests, plowing of meadows, damming of rivers and unrestricted hunting had taken their toll on wildlife.

To halt the decline in animal populations, laws that regulated hunting were established in many states in the early 1900s. This worked well for some animals, such as black bears and white-tailed deer, which gradually returned on their own to areas they had once inhabited.

In the case of animals that did not return on their own, wildlife biologists tried to raise them on game farms for release into the wild, or they captured wild animals in other states where they were established for release locally.

However, it was not enough to return these once-native animals to their historic territories. Their habitats – the land and water – also had to be restored. Loss of habitat is the primary reason most species of plants and animals become rare or endangered. So thousands of acres of forests were bought to establish state and federal forests,

The bald eagle was once the symbol of vanishing wildlife.

parks and wildlife refuges, and rivers that had been fouled with sewage and trash were gradually cleaned.

If animals were to be restored, they also had to be protected. In 1973, the federal Endangered Species Act was passed, making it a crime to hunt or otherwise harm many species of rare plants and animals. In addition, laws were passed in each state to protect plants or animals that were rare or endangered in that state.

State laws were also passed to protect wetland areas and water bodies, such as ponds, lakes, marshes and rivers. All living things depend on water for survival, and areas where land and water come together, such as the land surrounding a stream or river, are among the richest habitats for plants and animals. Red-winged blackbirds and marsh wrens nest in the cattails of pond edges. Beavers built their lodges in ponds and marshes using trees from surrounding land. Muskrats and river otters live much of their lives in water.

All these efforts have helped to restore both wildlife and wild lands. The bald eagle, nearly gone from Pennsylvania just thirty years ago, has become abundant enough that in 2014 it was taken off the state's list of endangered and threatened species.

Today in Pennsylvania, there are nearly 4,000 species of plants and animals considered native to the state. That does not even include the unknown number of invertebrate species – mainly insects.

Pennslvania's rare native species

Governments have set up a system to classify the health of populations of plants and animals. A species is considered endangered if it is on the verge of extinction in its natural range. It is threatened if it is likely to become endangered in the near future. It is on the federal endangered species list if it is endangered or threatened nationwide. It is on the state list if it is endangered or threatened in Pennsylvania even though it may be more common in other states.

Federal endangered species list includes these Pennsylvania natives:	State endangered species list includes:
Piping plover	All Pennsylvania species on the federal list plus:
Indiana bat	
Sheepnose mussel	Eastern mud turtle
Dwarf wedgemussel	Rough green snake
Red knot	Great egret
Northeastern bulrush	Peregrine falcon
Shortnose sturgeon	Northern flying squirrel
	Dragon's mouth

Protected Pennsylvania species as of 2015

	Endangered state list/ federal list	Threatened state list/ federal list
Mammals	3/2	3/0
Birds	17/1	3/1
Reptiles	5/0	1/1
Amphibians	5/0	2/0
Fish[1]	28/2	10/0
Invertebrates[2] not known	10/12	1/1
Plants	228/1	78/1
Total	296/18	98/4

[1] Includes inland freshwater fish only
[2] Includes butterflies, moths, beetles, dragonflies, crustaceans, mussels, snails, worms and sponges

Great spangled fritillary, a native species

Black-eyed Susan, a native species

Invaders

Just like many Americans, many plants can trace their roots to foreign soils. There are also animals found in Pennsylvania that have ties to other places in America or abroad.

A range of wildflowers, trees, insects, fish, birds and other life forms were brought to the region over the years by gardeners, naturalists, cooks and sportsmen who admired them elsewhere and released them – or allowed them to escape – into the wild here.

Others came as accidental hitchhikers on trucks, cars and airplanes and as stowaways on early sailing ships and ocean freighters. Still others arrived here on their own as part of the natural spread of their species into places where they could survive.

While the vast majority don't harm local ecosystems, a small percentage do. Called "invasives," these aggressive species invade and dominate the landscapes on which they alight, pushing out valued native species. Some naturalists estimate that only about two-thirds of the plants found in the state were here before Europeans arrived. However, only two or three dozen of these late arrivals are considered seriously invasive.

For instance, purple loosestrife was introduced from Europe as a garden flower and eventually escaped into the wild. It has strikingly beautiful flowers, but it can quickly take over a wetland, such as a

Oriental bittersweet, an invasive species introduced from Asia

marsh, pushing out native plant species, some of them important to nesting wetland birds.

European starlings, which were also natives of Europe, were released in Central Park in New York City in the early 1890's by Shakespeare fans who wanted to introduce to America the birds mentioned in the immortal writer's plays. They have since spread from coast to coast and are reducing the populations of some other birds. Starlings nest in tree cavities, taking away nesting sites from other cavity-nesters such as bluebirds, northern flickers and tree swallows.

Oriental bittersweet, introduced from Asia, is a fast-growing vine common to back yards that wraps around shrubs, trees and flowers, forming a thick tangle in which native species perish. In the fall it produces reddish-orange berries.

At least one species of harmful invader, the gypsy moth, was the result of a breeding experiment gone wrong. In 1869, hoping to start a silk business, a Massachusetts man tried to cross imported European gypsy moths with silk moths to create a more productive silkworm. The breeding attempt failed, but some of the moth larvae accidentally escaped. Gypsy moths have now spread through most of the East.

In some years, the gypsy moth caterpillars, which feed on tree leaves, have damaged more than a million acres of forests.

Purple loosestrife, an invasive species introduced from Europe

European starling, an invasive species introduced from Europe

Eastern screech owl

From northern cardinals and black-capped chickadees to great horned owls and bald eagles, the skies can seem nearly as populated with birds as the land is with people.

In fact, more populated. It's estimated there may be 20 times as many birds on Earth as humans. Walk through any field or forest in summer and you can easily believe it.

Some 220 species of birds nest each year in Pennsylvania, and as many as 200 other species are seen in the state. In all of North America, more than 800 species of birds have been reported. Worldwide, there may be more than 10,000 species of birds.

While large forests in the state attract their share of birds that seek isolation from humans, such as scarlet tanagers and hermit thrushes, cities and suburbs are home to birds that can easily live around people, such as American goldfinches and mourning doves.

Cedar waxwing

Nearly 150 million years ago, the first birds were appearing on Earth. But they may have looked more like winged lizards than modern birds. Birds, like most creatures, have changed, or evolved, over millions of years to have specialized features that help them survive by giving them an advantage over other living things. For most birds, the most important feature is the ability to fly and to live much of their lives above ground and in the air.

Even among birds, though, different species have evolved to have different features – variations in size, coloring, wing design or body shape. Because of these physical differences, some birds can find food in ways that others can't, or they can live and nest in places where others can't. As a result, more kinds of birds can survive than would be possible if all birds ate the same foods and lived and nested in the same places.

Birds born with a difference that gives them an advantage in finding food, such as a slightly longer beak, or an advantage in escaping predators, such as greater flying speed, are more likely to survive and have young. Therefore, their special features are more likely to be passed along to future generations.

Using this process, called natural selection, woodpeckers developed over millions of years to have strong, sharp beaks, long tongues and sharp claws. They use their claws to grab onto the sides of trees, and they use their beaks to peck nest holes in trees and to dig beneath bark to find insects. They use their long tongues to grab the insects they find.

Herons and egrets developed to have long, thin legs and long, sharp beaks. They use their legs to wade in the shallow water of lakes, streams and ponds. They can stand very still and wait for fish or tadpoles to pass by so that they can spear them with their beaks. To a fish, their thin legs may look like reeds.

Artists of the air

Birds do not merely fly. They also hover, soar and glide. If there are artists of the air, though, it may be the hummingbirds. They have made an art of functional flight.

Hummingbirds evolved to have strong wing muscles for a specific purpose. They can beat their wings very rapidly, on average 55 times each second, to hover at flowers like bumblebees, so that they can use their long beaks and tongues to sip the energy-rich nectar inside flowers.

A female ruby-throated hummingbird

A male ruby-throated hummingbird

In the Northeast, songbirds (a category that includes most backyard birds) usually nest and lay their eggs from April to June. For smaller birds, like the tufted titmouse, white-breasted nuthatch and tree swallow, the eggs take about two weeks to hatch, and the young stay in the nest two to three weeks before they take their first flights.

For larger birds, such as great horned owls and red-tailed hawks, the eggs may take nearly four weeks to hatch, and the young may remain in the nest four to six weeks.

Birds in the wild have varying life spans. Harsh weather, accidents, predators, disease and lack of food can take their toll. Most songbirds live only two to five years.

American robin

However, some larger birds, such as mallards and great blue herons, may live 20 years or more.

Birds are often specialists about food. Many species prefer one type of food. Canada geese like grass, eagles prefer fish, finches look for seeds and warblers like insects. However, most birds will eat more than one kind of food, and some will change their diet as the seasons change.

Generally, the larger a bird is, the faster it can fly. Finches and sparrows can reach top flying speeds of about 20 miles per hour, hawks can fly 30 to 40 miles per hour and geese, when pressed, can fly 60 miles per hour. Peregrine falcons, which hunt from the air, may be capable of the fastest flying speeds,

Backyard birds of Pennsylvania

Some birds will nest in residential areas. Shown for the birds below is: Nest construction – Average number of eggs per brood – Broods per season – Period during which eggs are laid – Time to hatching – Time to nestlings' first flights

Black-capped chickadee

A cavity in a standing dead tree lined with cottony fibers, fur, moss, hair, wool and feathers, usually 4 to 10 feet off the ground – Six to eight eggs – One brood – Mid-April to mid-July – About 12   days – About 16 days

Ruby-throated hummingbird

A small cup of plant down, fibers and spider silk built in the fork of a drooping limb, usually 10 to 20 feet off the ground – Two eggs – One or two broods – Mid-May to mid-July – About 16 days – About 19 days

Blue jay

A cup of rootlets, twigs, bark strips and leaves, usually in an evergreen tree 10 to 25 feet off the ground – Four or five eggs – One brood – Early April to early July – About 17 days – About 19 days

American robin

A deep cup made of grasses and weed stalks and shaped with mud in the fork of a tree branch, in shrubs or on a window ledge, usually 5 to 15 feet off the ground – Three or four eggs – Two, rarely three broods – Mid-March to mid-July – About 12 days – About 15 days

Downy woodpecker

A cavity in a living or dead tree, usually 20 to 30 feet off the ground – Four or five eggs – One brood – Early May to late June – About 12 days – About 21 days

Northern cardinal

A shallow cup of twigs, grasses, rootlets and vines in dense shrubs or thickets or in an evergreen tree, usually less than 10 feet off the ground – Three or four eggs – Two or three broods – Early April to early August – About 12 days – About 12 days

House sparrow

A cavity lined with grasses, weeds and feathers in a tree, on a building, on a billboard or in a birdhouse, usually 10 to 50 feet off the ground – Four to six eggs – Two or three broods – Early April to early August – About 12 days – About 16 days

Blue jay

Northern cardinal

up to 200 miles per hour. They dive rapidly on their prey, usually other birds in flight, killing them with the impact of the collision.

Feathers are a bird's clothing, keeping it warm in winter and giving it a colorful appearance. A robin may have 3,000 feathers. A tundra swan may have 25,000. However, most birds replace their feathers once or twice a year, a process called molting. The molt may take a month or more to complete.

Some birds, such as male American goldfinches, grow their brightest feathers for the breeding season, when they are trying to attract mates. Then they grow duller-colored feathers for the winter, which makes them less visible to predators.

Since it is most often the male's job to attract a female, in many species, including northern cardinals and American goldfinches, the male has brighter feather coloring than the female. With her plain coloring that often blends in with surrounding vegetation, the female does not draw as much attention to the nest and young.

Eggs
Some bird eggs shown their actual size

For many birds, the color and markings on their eggs have a purpose – to act as camouflage. For birds that build cup nests off the ground, such as cedar waxwings, the base color of their eggs is usually cream or buff and the markings may be streaks or speckles of brown. For birds, such as terns and gulls, that build nests on the ground, where predators are even more of a threat, the camouflage of streaks and speckles can be even more of a necessity. For birds that lay their eggs in darkened holes in trees, such as some owls, the color usually does not matter.

Canada goose

American crow

American robin

Great horned owl

Cedar waxwing

Yellow warbler

Ruby-throated hummingbird

A robin's nest

When a robin's nest shows up unexpectedly on a family's window sill in spring, it instantly becomes a made-to-order field experiment and one of the most compelling entries into the natural sciences that young kids will encounter.

Robins love the worms found in residential lawns, and they are not shy about building their nests near or on a home – or even on a back-yard deck, as in this pictured nest.

In Pennsylvania, robins lay their eggs from mid-March to mid-July in cup-shaped nests made of grasses and weed stalks plastered together with mud. The nests might be built in the fork of tree branches or on a protected flat surface of a home.

With their distinctive sky-blue coloring, the eggs hatch in about 12 days. The fragile hatchlings initially weigh about five or six grams. As a comparison, a penny weighs about 2.5 grams. But within two weeks, when they are ready to take their first flights, they will weigh nearly 70 grams, which is about the size of an adult robin. And that means they will have to eat a lot in a short time to gain weight so rapidly.

Robin eggs in a nest

Females do most of the incubation of the eggs, as is the case for most songbirds. But when it comes to feeding the young, both parents do their part.

In the first few days, the parents feed the young swallowed food that they bring back to their mouths, but about day three or four, the nestlings begin eating whole foods, mainly worms and insects.

Studies have found the parents bring food to the nest six to seven times an hour, on average, so that each nestling gets 35 to 40 feedings per day. The parents bring about 200 grams of food to the nest each day, or about seven ounces. That amounts to more than six pounds of wriggly crawly things fed to the young before they leave the nest.

When young robins finally take their first flight about two weeks after hatching, it is usually an awkward trip from the nest to the ground, as they are not yet capable of full flight. The parents may lead them to the protection of shrubs or undergrowth, and they will continue to feed them for several days until they develop the skills to survive on their own.

One day old

Three days old

Two weeks old

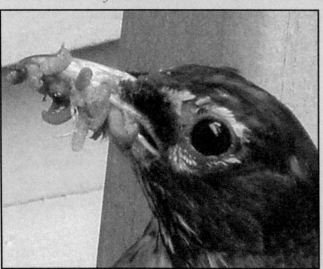

Food for the young

A feeding

Barn swallows build nests of mud and straw that are plastered to building structures.

The stick nests of bald eagles are built high in trees and are used year after year.

Killdeer nests are built on the ground in open areas.

Hummingbird nests are often held together with spider silk.

Nests

Different species of birds build different kinds of nests. One of the most common types is a cup-shaped nest built with grass or twigs in the fork created by two branches of a tree. However, birds also build nests on the ground, atop buildings, on billboards, in barns, on mountain ledges and in holes in trees.

House wrens have been known to build a nest in a discarded shoe, a tin can, a teapot or in the pocket of a shirt hanging on a clothesline.

The largest nests are those of eagles. Some are eight feet across, 15 feet deep, and weigh as much as a small automobile.

The smallest nests are those of hummingbirds. Some are less than an inch across and the eggs are the size of peas.

Pileated woodpeckers peck holes in trees for their nests.

Yellow warblers build nests of milkweed fibers, plant down, hemp and grasses.

The common loon builds its nest right by the water using grass, reeds, twigs and rushes. (It visits but does not nest in the state.)

Cedar waxwings at their nest

An eastern bluebird entering its nest in a tree hole

A mute swan on its nest in an urban pond

An American redstart checking its chicks

A northern cardinal feeding its young

American goldfinch

Tufted titmouse

Mourning dove

Great horned owl

Baltimore oriole

Songs and calls

The songs and calls of birds are their language. They use these sounds to tell other birds a territory is theirs, to attract a mate and to warn of danger, among other things. Their songs and calls may sound like this:

TUFTED TITMOUSE – "Peter! Peter! Peter!"
MOURNING DOVE – "Coo-ah, coo, coo, coo!"
GRAY CATBIRD – "Meow!"
OVENBIRD – "Teacher, teacher, teacher!"
GREAT HORNED OWL – "Who, who-who-who, who, who!"
EASTERN TOWHEE – "Drink your tea!"
BALTIMORE ORIOLE – "Hew-lee!"
WHITE-THROATED SPARROW – "Poor Sam Peabody, Peabody, Peabody!"
AMERICAN GOLDFINCH – "Potato chip!"
RED-WINGED BLACKBIRD – "Ok-a-lee!"
COMMON YELLOWTHROAT – "Witchity, witchity, witchity, witch!"
BLUE JAY – "Thief! Thief!" or "Jay! Jay!"

Migration

Because northern winters can be harsh and food can be scarce, many birds migrate in the fall to warmer regions in the southern United States and Central and South America. Then they return in the spring. However, some birds stay in Pennsylvania through the winter, including northern cardinals, blue jays and crows. Some birds of a species, like the eastern bluebird, will stay in the state through the winter, but most migrate south.

During the migration, birds may travel 200 to 300 miles a day. Some, like hawks and ducks, travel during daylight hours, but most smaller birds travel at night when they are less visible to predators.

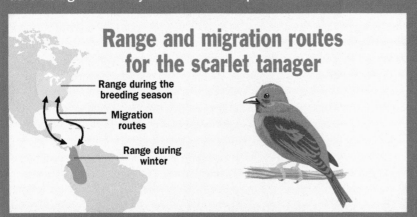

Range and migration routes for the scarlet tanager

Range during the breeding season

Migration routes

Range during winter

Typical migration times in Pennsylvania

	Arrival	Departure
Eastern bluebird	Early March	Oct.
Song sparrow	Late March, early April	Sept., Oct.
Eastern meadowlark	Mid-March, early April	Oct.
House wren	Late April, early May	Sept.
Broad-winged hawk	Late April	Mid-Sept.
Yellow warbler	Early May	Late July, early Aug.
Barn swallow	Late April, early May	Mid-Aug.
Baltimore oriole	Early May	Late Aug., early Sept.
Scarlet tanager	Mid-May	Sept.

CHECKLIST
Common birds

Red-winged blackbird 8 in. from tip of beak to tip of tail

☐

Eastern bluebird 6 in.

☐

Northern cardinal 8 in.

M   F

☐

Black-capped chickadee 5 in.

☐

American crow 19 in.

☐

Mourning dove 12 in. ☐

Rock dove (pigeon) 13 in. ☐

House finch 6 in. ☐

Northern flicker 13 in. ☐

American goldfinch 5 in. M F ☐

Canada goose 38 in. ☐

Common grackle 12 in. ☐

Ring-billed gull 19 in. ☐

Great blue heron 48 in. ☐

Ruby-throated hummingbird 3 in. M F ☐

Blue jay 12 in. ☐

Dark-eyed junco 6 in. ☐

Belted kingfisher 13 in. ☐

Mallard 23 in. ☐

Northern mockingbird 11 in. ☐

White-breasted nuthatch 6 in. ☐

Baltimore oriole 8 in. ☐

American robin 10 in. ☐

House sparrow 6 in. ☐

Song sparrow 6 in. ☐

White-throated sparrow 7 in. ☐

European starling 8 in. ☐

Tree swallow 6 in. ☐

Tufted titmouse 6 in. ☐

Cedar waxwing 7 in. ☐

Downy woodpecker 6 in. ☐

House wren 5 in. ☐

Raptors are the rulers of the skies. Eagles, hawks, falcons, owls and other birds of prey (as raptors are also called) have features that are standards of excellence in the animal world. Consider the sharp eyes of an eagle, the keen hearing of an owl or the breathtaking speed of a diving falcon. Throughout the Northeast, the populations of many species of these raptors are soaring. It has become common now to see a red-tailed hawk sitting high on a roadside tree or an American kestrel perched on a telephone wire.

In the 1960s, though, some species of raptors were headed for extinction. A widely used chemical pesticide, DDT, caused them to lay eggs with no shells or shells so thin that they broke. DDT was banned in 1972, and almost immediately, populations of birds of prey began to recover.

In 1963, there were only 487 known pairs of nesting eagles in the lower 48 states. Today, there are nearly 10,000 pairs. In 1983, there were only three active eagle nests in Pennsylvania, all in Crawford County in the northeast-

Great horned owl

ern part of the state. Today, there are more than 250 active eagle nests in the state, with nests present in at least 59 of the 67 counties.

Peregrine falcons, whose population was wiped out in the East by DDT, also recovered, partially through the reintroduction of falcons into Pennsylvania from other regions where they still could be found during the 1970s. And in the spring of 2015, there were 43 known nest sites in the state.

Many raptors have also been helped by changes in how land is used. A little more than a century ago, farm fields covered much of the Northeast, but many of those farms have been abandoned, and the fields have grown back to forests.

Today, the region has a rich mixture of landscapes – patches of forests that are alongside patches of meadows, farm fields or residential neighborhoods. And this change has helped many species, especially those that like to perch in trees on the edges of forests and hunt small animals in open areas, such as Cooper's hawks and red-tailed hawks.

However, a few species have been hurt by the changes, especially barn owls and northern harriers, which prefer farms and fields.

Red-tailed hawk nest

American kestrel

Broad-winged hawk

Raptors in flight

(Owls, which are generally nocturnal, are not usually seen in flight.)

Peregrine falcon

Bald eagle

Broad-winged hawk

American kestrel

Cooper's hawk

Northern harrier

Golden eagle

Northern goshawk

Red-shouldered hawk

Barn owl

Osprey

Snowy owl

Bald eagle

Sunset at the Birdsboro Reservoir in Berks County

2–3 in. long

Graceful and elegant, deer are animals that a child might imagine in dreams. To see a doe and its fawns grazing in a meadow in the early morning light or an antlered buck nibbling tree seedlings in a sunlit forest is to see something that one does not easily forget.

But white-tailed deer graze in places other than meadows and forests. They graze in farm fields, in vegetable gardens and on ornamental shrubs. They also wander across well-traveled highways in search of still more grazing opportunities. And where deer go, Lyme disease, which is carried by deer ticks and can affect humans, may follow. So not everyone loves to see whitetails.

A century ago, deer were rare in much of the Northeast. Sightings of them were often reported in newspapers. It is estimated that in 1900, there were only about 5,000 whitetails in all of Pennsylvania. In an effort to replenish the deer population, limits were put on hunting them and hundreds of deer were brought in from other states where they were more plentiful. It worked – perhaps too well.

Today there may be 1.6 million deer spread throughout the state, from rural forests to thickly settled, suburban neighborhoods. As a consequence, Pennsylvania leads all states nearly every year in the number of deer-vehicle collisions (an many as 100,000 a year) as well as in the number of new Lyme disease cases.

White-tailed deer may gather in dense stands of evergreens, called "deer yards," in severe winters with deep snow. Snow depths are usually less there, and deer can browse on the branches. The trees also give them protection from wind, cold temperatures and blowing snow.

In the wild, white-tailed deer may live seven years or more. If food is abundant, a doe may breed when she is only six months old and have a fawn around her first birthday in the spring. However, most does first breed as yearlings. In later years, she may have one to three fawns each spring, which means deer populations can grow rapidly.

As with many animals, the males are larger than the females. Bucks usually weigh 100 to 200 pounds, and does may reach 70 to 150 pounds. Bucks grow antlers each year, usually starting in April, and shed them in winter, late December to February.

There is a saying that when a leaf falls in the forest, an eagle will see it, a bear will smell it and a deer will hear it. Gifted with extremely sharp hearing, white-tailed deer are tough and hardy survivors. They use their hearing and

White-tailed deer

swiftness afoot – they can run at speeds of 35 miles per hour – to avoid the few predators they have, and they can live close to residential neighborhoods while rarely being seen.

Deer prefer areas where there are forests mixed with clearings, wetlands, abandoned pastures or active farms. They are vegetarians, favoring the buds and twigs of young trees in the cold months, and grasses, fresh leaves and nuts, such as acorns, in the warm months.

A disease that affects both deer and elk, called chronic wasting disease, was first detected in a small number of deer in Pennsylvania in 2012. First identified in Colorado in 1967, the illness attacks the central nervous system and causes the animal to lose so much weight that it dies. There is no known treatment. So far, there is no evidence that eating infected meat affects humans.

Elk in Pennsylvania

They arrived by train in 1913, disembarking in Clinton and Clearfield counties to settle permanently in Pennsylvania. These 50 elk, which had been collected in Yellowstone National Park, where elk were still plentiful, were the first shipment in an effort to re-establish a once-native animal that had vanished from the state, perhaps by the 1870's, a victim of over-hunting and the loss of its favored habitat.

Today, there may be 900 elk in the state, most living on state-owned lands in north-central Pennsylvania, in the region where the 1913 group was let loose.

Elk, or wapati (which is believed to be the Shawnee Indian term for elk), are a kind of deer, as are moose. And like its male cousins, elk bulls grow antlers that are shed in late winter each year and regrown over the spring and summer. An elk's antlers can stand nearly four feet above its head and weigh up to 40 pounds.

Elk are larger than whitetails, weighing, on average, five times as much. The moose, which is not found in Pennsylvania, is the largest of the three. A bull moose may weigh ten times as much as a deer buck.

White-tailed deer (buck)
Height: 3 feet at the shoulder
Weight: 100 to 200 pounds

Elk (bull)
Height: 4.5 feet at the shoulder
Weight: 700 to 800 pounds

Moose (bull)
Height: 6 feet at the shoulder
Weight: 1,300 to 1,500 pounds

A male elk, right, called a bull, grazing with a female, called a cow

6–7 in. long

What is there about a bear that has such a dramatic effect on people? From affection for panda bears to fear of grizzly bears, human emotions run high at the sight of them. Indeed, to come face-to-face with a bear is perhaps the ultimate wildlife experience of the forest – and increasingly of the suburbs.

There may now be 18,000 black bears roaming Pennsylvania. Found mainly in large forested regions, bears have nevertheless been sighted in every county in the state. With no major predators except humans, a bear can wander just about anywhere it wants, and many have discovered the food resources of residential areas, including bird feeders and trash cans.

Black bears were nearly gone from Pennsylvania by 1900, when nearly 70 percent of the state's land was unforested, mainly for use in farming. But restrictions were placed on bear hunting and the population rebounded and spread.

Black bears have long suffered from a public relations problem created by the larger, more aggressive grizzly bears, which are not found in the eastern United States. The difference is that eastern black bears rarely attack human beings. Usually, they are fairly docile creatures and won't attack unless they are defending a cub or have been surprised.

Black bears can live 25 years or more. Males typically weigh 130 to 500 pounds, while females may weigh 100 to 300 pounds.

Bears eat little meat, focusing instead on plant matter that is available through the seasons, from skunk cabbage and grasses in the spring to apples and acorns in the fall, when they are trying to fatten up for their winter sleep.

Black bears are not true hibernators. While a bear's heart rate drops dramatically during the winter dormancy, its temperature does not. Female bears give birth to young in the den, typically in January, so a high body temperature is necessary for them to care for the young.

However, their taste for honey is well known. Black bears have an extraordinary sense of smell that they use to locate food and detect danger.

In Pennsylvania, bears are usually active in daytime during the spring and fall, but they are more active during the hours around dawn and dusk in summer.

Females give birth during their winter dormancy every other year, usually having two to four cubs.

Black bear

Black bear raiding a bird feeder

To see a red fox in the wild is memorable. True to its reputation, a fox is as sly as ... well, a fox. Very secretive, it can live close to residential areas but almost never be seen. It's also an impressive looking animal with its silky red fur and sharp features.

Red fox track, 2 in. long

Red foxes are not animals primarily of the deep forest. They prefer farmland or forest edges. The suburbs, with its mix of fields and small wood lots, can also be an excellent habitat for them.

By contrast, gray foxes, which are less common in Pennsylvania, usually live in forested or brushy areas.

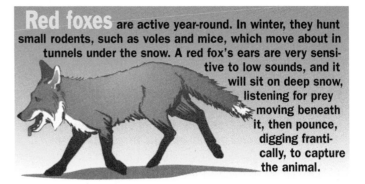

Red foxes are active year-round. In winter, they hunt small rodents, such as voles and mice, which move about in tunnels under the snow. A red fox's ears are very sensitive to low sounds, and it will sit on deep snow, listening for prey moving beneath it, then pounce, digging frantically, to capture the animal.

Gray fox

Both red and gray foxes are smaller than most people expect, usually weighing just seven to 14 pounds, about as much as a large house cat. They have varied diets, eating what animals and plants are available through the seasons, including rabbits, rodents, birds, insects, snakes, turtles, apples and berries.

Very often, foxes will not eat what they capture, especially if they are full. Instead, they will bury the prey for a later meal.

Vixens, or female foxes, bear their young in dens in late March or early April. They may have four to seven kits, which begin to spend time outside the den in early May. By late summer, the young will leave their parents' care permanently.

Unlike red foxes, gray foxes, which have sharp claws, will climb trees to reach potential food, such as bird eggs. They will also scamper up a tree to escape danger. While a red fox den is nearly always underground, a gray fox will sometimes den off the ground in a hollow tree.

Red and gray foxes are found throughout the state.

Red fox

Eastern coyote track, 2.5 in. long

"Ravenous cruell creatures." That was the description our Colonial ancestors gave of wolves.

Because of repeated attacks on their livestock, European settlers offered bounties for these stealthy predators, and gradually a predator that had been at the top of the food chain in the Northeast was eliminated. The last wild wolf known to be living in Pennsylvania was killed in the 1890's.

It has long been assumed as fact that wolves no longer reside in the state. However, it now seems that assumption has been wrong – well, half wrong.

The wolf may again be on the prowl in Pennsylvania in the form of the eastern coyote. Genetic tests of tissue samples of coyotes in the Northeast have found evidence of wolf genes in the mix, sometimes to the point that the animals tested were more wolf than coyote.

Today, there are two wolves native to North America – the gray wolf, also called the timber wolf, which is found mainly in Canada, and the smaller, and more rare, red wolf. Only a few hundred red wolves remain and they live mainly in captivity in zoos and breeding facilities.

The eastern wolf, found mainly in eastern Canada, was thought to be a subspecies of gray wolf. However, some now believe it too may be a distinct species of wolf.

There had long been the suspicion that the eastern coyote had some wolf in it. It was known that coyotes, which were not found east of the Mississippi River

Eastern coyote

before 1900, migrated up into Canada from the western United States, crossed north of the Great Lakes and appeared in Pennsylvania in the 1930s. However, this newly arrived coyote in the Northeast was noticeably larger than the coyote of the West. Did coyotes and wolves interbreed in Canada?

It took the rise of biotechnology and genetic analysis in the 1990s to confirm the suspicions. Indeed, researchers found wolf genes in eastern coyotes – mainly those of the eastern and gray wolves.

It is known that canids – members of the family that includes wolves, coyotes, foxes and dogs – will interbreed when mating partners of their own kind become scarce. And when so many wolves were killed when the continent was being colonized, wolfs apparently did mate with coyotes, producing our eastern coyote.

All this has thrown views of North America's "wild dogs" – wolves and coyotes – into a state of confusion. Some say the eastern coyote should more properly be called something else, perhaps even a new type of wolf.

Eastern coyotes are now found throughout the state, but they are more common in the northern half.

Eastern coyote
Length: 4-5 ft. (nose to tail)
Weight: 43 lbs. (average male)

Western coyote
Length: 3.5-4.5 ft. (nose to tail)
Weight: 33 lbs. (average male)

Gray wolf
Length: 5-6 ft. (nose to tail)
Weight: 80 lbs. (average male)

Red wolf
Length: 4.5-5.5 ft. (nose to tail)
Weight: 65 lbs. (average male)

Eastern coyotes

The eastern coyote can look a lot like a German shepherd dog. However, coyotes are usually smaller and leaner. Also, a coyote's tail will usually be carried lower to the ground than that of a shepherd.

Eastern coyotes usually weigh 30 to 50 pounds, more than western coyotes. They are often grayish tan in color, but their coats can also be blond, red or even charcoal black. Strong swimmers, good jumpers (up to 15 feet) and swift runners (speeds of nearly 40 miles per hour have been observed), Eastern coyotes are also able hunters, with superior sight, smell and hearing.

Coyotes, which are most active from dusk to dawn, are usually no threat to humans. They are expert scavengers, eating almost anything edible from the bottom of the food chain to the top – from berries, fruits and insects to mice, rabbits and, yes, even the occasional house cat.

Eastern coyotes hunt alone, in mated pairs or in family groups. Coyotes are famous for their howls in the night, a form of communication for them. They will even howl in response to sirens, car alarms and train whistles. But often, the howling is meant to tell non-family members to stay out of the territory.

5 in. long

Beavers are nature's engineers, creating dams and lodges that are marvels of construction.

However, their dense, water-resistant fur has always been highly prized for clothing, and uncontrolled trapping during Colonial times eliminated beavers from much of the Northeast. By 1900, they were virtually gone from Pennsylvania.

However, regulated trapping and the reintroduction of beavers from other states in the early 1900s allowed this industrious mammal to repopulate Pennsylvania. Today, beaver colonies are found throughout the state, although they are most common in the northern rural counties where there is a constant source of water.

But as the beaver population has grown and spread, these animals have caused problems in some residential areas. Having a beaver pond in the neighborhood is a delight to many people, but a nuisance to others as the ponds can sometimes flood back yards, roads and wells.

From their protruding, orange-tinted, buck teeth to the end of their paddlelike tails, adult beavers are two to three feet long, and they can weigh 30 to 70 pounds. They use their broad, flat tails as rudders when they swim, but they also use them to warn other beavers of danger by loudly slapping them on the water's surface. A beaver can stay submerged in water for up to 15 minutes.

Beavers use trees as their prime building material, and they also feed on the bark. Their favorite trees include poplars, birches, maples, willows and alders. They usually fell trees at night and are able to gnaw their way through a willow that is five inches in diameter in just a few minutes.

Beavers dam streams to create ponds. They spend much

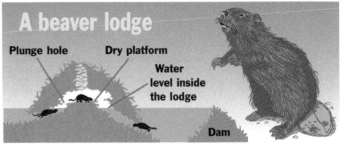

A beaver lodge

Plunge hole　Dry platform　Water level inside the lodge　Dam

of their lives in water, which protects them from predators. They begin by laying branches and twigs across a stream that may be in a slight valley. Then they pack down this material with mud, extending the dam's width and height as they go.

Their trademark lodge can sit near the shore of a pond or right in the middle, surrounded by water. The roof of the lodge is made of twigs and mud. Inside, it has a dry platform just above the water line with one or two plunge holes for an entrance and exit. For protection, beavers almost always work on dams and lodges at night.

In winter, beavers do not hibernate. They remain awake in the lodge, occasionally going out beneath the ice to retrieve tree branches they've stored in the water nearby for a meal. A family of beavers can eat a ton of bark in winter. Beavers take a mate for life. The young, called kits, are born in May or June and will take to the water within hours.

Once they dam a stream, a family of beavers typically stays in that location five or six years, expanding the system of dams and ponds as the family grows. Several generations of the family can make up the colony.

A beaver lodge

A beaver and a freshly gnawed tree

A beaver's strong orange teeth

1.5 in. long

A blood-curdling scream deep in the night brings frantic calls to local police, but no evidence of any kind of crime is found.

The mystery is no mystery to wildlife biologists. They know that in the spring what may sound like a crime of passion is in many cases a bobcat in the heat of passion during mating. Its scream is similar in pitch to that of a human female.

Sly. Secretive. Solitary. Bobcats are a rare sight even in areas where they might be common. Bobcats may look like house cats who have done some serious weight training, but the two have quite different lineages. House cats are descended from the first cats domesticated about 4,000 years ago, when Egyptians took the African wild cat and tamed it to keep granaries free of rodents. Later, they were bred as pets.

Bobcats, on the other hand, derive from the Eurasian lynx that took the opportunity perhaps two million years ago to migrate from Asia to North America across a periodic land bridge that emerged across the Bering Strait during ice ages. The lynx learned to live in snow-free regions of the continent, including Mexico, and eventually evolved into the more muscular bobcat.

Visually, what distinguishes the bobcat from most house cats is its tail. It's a bobbed, or shortened, version, about six inches long, of what one usually sees on the house cat. The way a bobcat carries itself can also tell you this is not a house cat on the prowl. They have a certain confidence, a sense that they are only predator and not prey

The size of the bobcat population in Pennsylvania was estimated in 2000 to be about 3,500 animals, with most living in the mountains and deep forests.

Typically, bobcats are 28 to 47 inches long and weigh 15 to 35 pounds. Females have one litter per year of one to four kittens born in spring. The young are raised entirely by the female as the male takes no part in family life.

Bobcats, which live about 12 years on average in the wild, are active before dusk to just after dawn in summer and also during the daytime in winter. Their dens are usually formed in a rock crevice, cave, brush pile or hol-

Bobcats will capture an animal, such as a young or sick deer, in winter and eat until they are full. Then they will cover the carcass with snow to return to later for a meal.

low log and then lined with dried grass, leaves, moss, and other soft vegetation. They are active throughout the year.

Their diet consists of rabbits and other small mammals in summer, and in winter, snowshoe hares, if they can find them, and occasionally sick or weakened deer. For younger bobcats, starvation is the number one killer.

Bobcat

Black bear cubs explore a tree

A deer buck and doe at sunset

A bobcat mother and her kitten out for a walk

A raccoon hunts in a wetland

A young porcupine practices climbing

Red fox kits emerge from their den

Pittsburgh skyline with the Monongahela River in the foreground and the Allegheny River in the distance

Bald eagles, Bengal tigers and giant pandas are all darlings of science and the media, sharing a charisma factor. Bats, along with cockroaches and rats, suffer from a "yuck" factor and are largely shunned.

However, there would be a world of difference in a world without bats. They eat an amazing number of insects, primarily moths and beetles. It's estimated that a single bat can eat 600 insects in an hour (although contrary to popular belief, mosquitoes are not a principal part of a bat's diet).

We may soon experience a world without bats, though. In Pennsylvania and elsewhere bats are dying in high numbers. A fungus, called white nose syndrome, is threatening bats throughout the eastern United States and Canada, causing devastating losses of their populations.

Once the most common bat in the United States, the little brown bat is now being considered for the endangered species list. It's estimated that nearly 95 percent of little brown bats in the eastern half of the country have died from white nose syndrome.

First identified in New York in 2006 and in Pennsylvania in 2008, the disease has spread rapidly. As of 2014, it had been detected in bats in 25 states in the eastern United States and in five Canadian provinces. Some scientists believe the little brown bat may vanish from the country entirely in a little more than a decade.

Often described as "flying mice," bats are, in fact, not rodents. They are mammals. There are nine species of bats that are native to Pennsylvania, including the two most common, the little and big brown bats. Two other species are occasionally visitors to the state.

The little and big brown bats, not surprisingly, are both brown and both have soft fur covering their bodies. The little brown bat has about a 10-inch wingspan while the big brown bat measures 13 inches wing tip to wing tip. Typically, a little brown bat weighs just a quarter of an ounce, about the weight of a quarter, compared to two-thirds of an ounce for the big brown bat.

Prior to the current bat illness, the little brown bat was the species most people crossed paths with, as it has a habit of slipping into houses through impossibly small gaps and holes. In summer, during the day. the little brown may roost, or rest, in attics, behind window shut-

A bat's radar

Bats have fairly good eyesight, but in the dark, eyes have little value. So bats have evolved to have a kind of radar that depends on high-frequency sounds that humans can't hear. As a bat flies, it emits short cries at this frequency. The waves of sound go out and when they hit something solid, such as a tree branch or a flying insect, they bounce off the object and return to the bat, which picks up the returning sound with its ears. The bat's brain analyzes the time the sound wave took to return, whether it is an irregular signal and what its pitch is to determine how far ahead the object is. These things will tell it whether it is a flying insect, and whether or not it is moving toward the bat or away from it. For instance, the fluttering wings of a moth might produce a fluttering echo wave.

Echo wave

ters and in other dark, hot, sheltered places.

While bats can be carriers of rabies, they are not usually infected. But health officials warn people never to handle a bat with bare hands and to treat any bat flying in the living quarters of a home as if it has rabies. The best method to get rid of it is to open a window and leave the room.

The bats in Pennsylvania have two main wintering strategies. The hoary, red and silver-haired bats migrate south with the birds. All of the others go to mines and caves in September and October to hibernate.

Why mines and caves? In winter, the temperature and humidity in these chambers can be considerably higher than conditions outside. During hibernation for little brown bats, breathing slows to one breath every five minutes and body temperature drops until it is only slightly higher than the surrounding air. Groups of them form tight clusters, hanging from the ceiling by their feet.

It is in these hibernating chambers that white nose syndrome, a fungus, spreads within these large groups of sleeping bats, killing most of the infected bats through the winter.

Little brown bat

1.5 in. long

"Once bitten, twice shy" certainly applies to those who have encountered striped skunks. Sprayed just once with a striped skunk's unmistakable scent, and you will never want to repeat the experience.

Skunks have one of the more effective defenses against predators on the planet. Their scent spray, which can travel 15 feet in the air if the wind is right, is so strong it can temporarily blind a victim.

Their bold coloring – black with a white stripe running down their back – makes the lesson that much more unforgettable. "If you see another animal that looks like this, stay clear," the distinctive colors seem to say.

Skunks, which usually weigh three to 12 pounds, eat almost anything, including snails, small rodents, birds' eggs, fruit, grain, nuts, berries and garbage.

Typically, they are nocturnal animals, and a sign one has been in your yard is shallow holes dug in the ground. A skunk was likely digging for grubs or worms during the night.

Skunks are usually active earlier in the spring than other mammals. They might spend the winter under a stump, in a stone wall, in an abandoned burrow or even under a house. But in mid-February the breeding season starts. Males may emerge from their dens and travel in search of females even when there is snow on the ground. This is a time of year when wandering males often cross paths with dogs, coyotes and other non-hibernating species. For many people, the pungent odor that results from such a

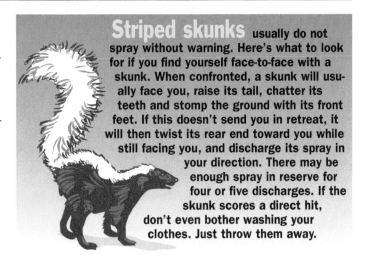

Striped skunks usually do not spray without warning. Here's what to look for if you find yourself face-to-face with a skunk. When confronted, a skunk will usually face you, raise its tail, chatter its teeth and stomp the ground with its front feet. If this doesn't send you in retreat, it will then twist its rear end toward you while still facing you, and discharge its spray in your direction. There may be enough spray in reserve for four or five discharges. If the skunk scores a direct hit, don't even bother washing your clothes. Just throw them away.

meeting is one of the first signs that spring is on the way.

The scent itself is composed of a compound containing sulfur that is secreted by the anal glands. In fact, skunks can be relatively gentle animals and typically will not discharge their scent except as a last resort.

Home remedies to remove the scent from skunk-doused pets often mention a bath in tomato juice, beer or vinegar. But studies have found a more effective mixture is one quart of 3 percent hydrogen peroxide (which can be obtained from a drug store), a quarter cup of baking soda (sodium bicarbonate) and a teaspoon of liquid detergent. This mixture chemically neutralizes the scent rather than just covering it up.

Striped skunk

Gray squirrel track, 2.5 in. long

The squirrel is one of nature's bankers, wisely putting away resources for the future. Each fall, eastern gray squirrels go about collecting nuts (their favorite food) and burying them one by one for the time when the snow falls and their sources of food disappear.

When winter arrives, they can find their stores, or caches, of acorns and other nuts beneath a foot of snow, not by remembering where they buried them but by smelling them.

Squirrels are rodents, as are chipmunks, woodchucks, beavers and mice. Most rodents hoard or store food to some extent.

In Pennsylvania, eastern gray squirrels are the squirrels most commonly seen by people. With their quickness and ability to move acrobatically through the treetops, gray squirrels live easily around people and can thrive in residential areas with trees.

Red squirrels are common in many areas with evergreen trees, including residential neighborhoods. But northern and southern flying squirrels, which are also native to the state, are usually found in more remote forested areas.

The fox squirrel, which is larger than the gray squirrel, is found mainly in western and southern counties in the state.

Gray squirrels may have two litters each year, usually of two to four young per lit-

Flying squirrels don't actually fly and they don't have wings. They spread open flaps of loose skin between their front and rear legs and glide through the air. Glides of up to 50 yards have been observed.

ter. One litter may be born in March and another in July or August. The spring litter is born in a tree den, which is a hollow cavity in a live tree also used during winter for shelter. The summer litter may be born in a large nest built high in a tree using leaves and sticks.

From its nose to the tip of its tail, the gray squirrel is typically 18 to 20 inches long, while the red squirrel is 12 to 14 inches long and the two flying squirrels are 9 to 11 inches in length. The fox squirrel may be 21 inches long, including a tail that is often 10 inches long. A fox squirrel may weigh 2 pounds, compared to 1.5 pounds for a gray squirrel.

Although squirrels may stay in their dens for several days at a time when the weather is harsh in winter, they are active throughout the year.

Northern flying squirrel

Fox squirrel

Eastern gray squirrel

Red squirrel

Yes, there are moose and bears and deer in the woods, but it is the small critters that really populate the great outdoors.

In a square mile of good habitat, there might be a few red foxes, a few dozen striped skunks, a few hundred eastern chipmunks and a few thousand short-tailed shrews. That's not to mention the brown bats, red squirrels, gray foxes, beavers, porcupines, cottontails, weasels, coyotes, woodchucks, otters, fishers, bobcats, muskrats, rabbits, voles, mice, minks and moles.

But if there are that many mammals out there, why do we see so few? The reason is that many species are secretive and have habits that help them avoid predators (and human beings). For instance, many are nocturnal – active mainly at night. They sleep and rest in hidden places by day but then go out foraging for food once the sun goes down.

With so many animals sharing the same land, many have evolved to be specialists in where they live and what they eat so that they can survive despite the competition.

Muskrats, beavers and river otters spend much of their life in water. Shrews, moles and voles spend much of their life underground or under the cover of thick grasses or leaves. Squirrels and porcupines spend most of their time in trees. And bats fly.

Porcupines, rabbits and woodchucks are vegetarians. Moles eat worms and ground insects. Bats eat flying insects. Skunks, raccoons and opossums will eat almost anything that can be eaten, including garbage.

Even among mammals, though, there are predators and prey, roles that are often determined by size. The shrews, mice, voles and moles tend to be the

Eastern cottontail

Porcupine

prey, while the foxes, coyotes and bobcats are often the predators.

When approached by a predator, prey animals will use their specialized defenses. The opossum can act as if it's dead. The eastern cottontail can leap 15 feet. Snowshoe hares, which turn white in winter and blend with the snow, can run 27 miles per hour. Striped skunks can spray a fluid that smells terrible and can cause temporary blindness. Porcupines can protect themselves with their coat of sharp quills – as many as 30,000 needle-like spines.

In winter, many mammals either hibernate or become inactive in their dens or burrows. But some, such as coyotes and foxes, are active throughout the year.

It is not just the deep woods and rural meadows that are home to so many mammals. Even urban and suburban neighborhoods can be teeming with wild animals, and not just mammals. Many snakes, birds and insects also call this kind of habitat home.

Because many mammals, such as opossums and skunks, are nocturnal, you rarely see them in residential areas. But they're there. Ask any policeman who drives a cruiser on the midnight to dawn shift.

In fact, when an area is transformed from a rural to a residential landscape, some animals find life easier. They learn to adapt their diet to the new opportunities these neighborhoods can offer, such as bird feeders, vegetable gardens, trash cans, and healthy lawns and shrubs.

Many animals, such as squirrels, chipmunks and a variety of birds, live comfortably with people and are often in sight. Other animals, including garter snakes and woodchucks, live easily among people but prefer to stay out of sight as much as possible.

Raccoon

River otter

Opossum

Woodchuck (groundhog)

Muskrat

Chipmunk

River otter

35 to 54 in. – 12 to 20 lbs. – One litter per year of two to four young born March or April – Active mainly from dusk to after dawn, but can be active any time of day – Eats primarily fish, but also frogs, turtles and aquatic insects – Dens in a rock crevice, under a fallen tree, in an abandoned beaver lodge or muskrat house or in thickets beside water – Active throughout the year

3 in. wide

Opossum

26 to 35 in. – 5 to 9 lbs. – One or two litters per year usually of six to nine young, but sometimes as many as 20, born February to July – Active mainly at night – Eats nearly anything, including insects, fruits, nuts, carrion and garbage – Dens in an abandoned burrow, tree cavity, brush pile or thicket – Becomes less active in winter, but does not hibernate

2.5 in. wide

Porcupine

24 to 30 in. – 8 to 20 lbs. – One litter per year of one young born April to June – Active mainly at night – A strict vegetarian, it eats grasses, twigs, buds and bark – Most den in winter in rocky fissures, hollow trees and logs or vacant buildings, such as sheds, but they may emerge to feed during the day – Active throughout the year

3 in. long

Eastern cottontail

14 to 18 in. – 2 to 4 lbs. – Three to seven litters per year of three to six young – Most active dusk to just after dawn – Eats mainly plant matter, such as grasses and herbs, in summer, and bark, twigs and buds in winter – Finds shelter year-round in brush piles, stone walls or abandoned dens and burrows – Active year-round

3-4 in. long

3.5 in. long

Muskrat

2 in. long

19 to 25 in. (half its length is its tail) – 1 to 4 lbs. – Two to four litters per year of three to eight young born April to September – Active mainly at night, but may be seen during the day – Eats mainly aquatic plants, such as cattails and pond weeds – Constructs a lodge of weeds on water or digs a den in a stream bank – Active throughout the year

Raccoon

22 to 34 in. – 15 to 25 lbs. – One litter per year of three to seven young born April or May – Active mainly at night – Eats everything from small animals and birds to grains and garbage – Dens in a ground cavity, hollow tree, hollow log, pile of rubbish, attic or chimney – Spends winters in its den, but does not hibernate

Chipmunk

1 in. long

8 to 10 in. – 2 to 5 oz. – One litter per year of three to five young born in April or May – Active all times of day – Eats mainly seeds, fruits, nuts and insects – Dens underground in a series of tunnels – Hibernates in winter, but may become active for short periods

Woodchuck

2 in. long

19 to 27 in. – 4 to 14 lbs. – One litter per year of four to six young born early April to mid-May – Active by day, especially in the early morning and late afternoon – Eats mainly plants, such as clover and grasses – Dens in a series of underground tunnels ending in a chamber with a grass nest – Hibernates in winter

Roadkill is one of those grisly facts of life.

But to many naturalists, what lies in repose by the roadside can be a source of valuable information if not general fascination. For instance, why do some animals regularly end up as highway fatalities and others do not? As agile as squirrels are, why are they so often victimized? And how is it that fragile little sparrows, which spend so much time down on the roadway, don't even make the list of top 10 road-killed birds (yes, there is such a list), yet the more reclusive gray catbird is right up there among the leaders?

It is unfortunate, but the carnage on the nation's highways is breathtaking. By some estimates, as many as a million animals meet their demise in traffic daily. It may be moose in Maine, white-tailed deer in Pennsylvania, armadillos in Arizona, rattlesnakes in Texas or whiptail lizards in New Mexico.

In Pennsylvania alone, it's estimated 100,000 deer a year are killed in collisions with vehicles.

One survey by a Massachusetts naturalist chronicled roadkill, some 4,000 dead animals, the researcher spotted along the state's highways from 1985 to 1991. To no one's surprise, gray squirrels topped the list. Among birds, pigeons, which often nest beneath overpasses, were the number one victim. Part of the explanation for why squirrels so frequently expire on highways is the sheer number of squirrels. However, part of the explanation may also be the strange mind of the squirrel. It may go half way across a road, see a car, and dart back right under the wheels.

Among road-killed mammals in the survey (which did not include domestic cats and dogs), the numbers went like this: 1,488 gray squirrels, 445 opossums, 270 raccoons (before the current raccoon rabies outbreak that began in New York in 1990), 196 striped skunks, 194 cottontail rabbits, 111 woodchucks, 100 chipmunks, 77 muskrats and 56 red squirrels. The same study in Pennsylvania would likely find deer higher on the list.

Although residents of the air, birds nearly always end their lives on the ground, many of them on the highways of America. (A few are plucked from midair by predatory birds.) The survey included 1,000 road-killed birds, and it found 166 pigeons, 163 robins (they tend to fly low across the road), 80 starlings, 75 blue jays, 65 gray catbirds (another low flyer), 62 grackles, 41 flickers (a fatal attraction to ants), 33 eastern screech owls (they chase flying insects), 32 crows (they scavenge roadkill) and 31 mourning doves. Why aren't chipping and house sparrows, ever-present on road shoulders, high on the list? Perhaps it is just good instincts and maneuverability.

Then there is the question of what becomes of roadkill. In many communities, highway crews remove the remains. But often, nature supplies its own cleanup crew, from microscopic bacteria to scavengers as large as turkey vultures. Crows, raccoons, coyotes and even an occasional red-tailed hawk also feed in traffic.

A crow feeds on a road-killed squirrel

Turkey vultures

The undertakers of the air, turkey vultures may spend most of the day soaring in the sky, searching for food using both their eyesight and keen sense of smell. The odor they are seeking is ethyl mercaptan, a chemical created by rotting meat.

However, circling vultures do not necessarily mean there is a dead animal below. The birds could just be traveling higher on a thermal, a rising column of warm air. Riding these thermals, a turkey vulture with a wingspan of about six feet can glide for as much as six hours without once beating its wings.

Turkey vultures feed on dead animals. They will not circle or stalk a dying or healthy animal. However, plant matter is also a part of their diet.

The males and females are identical in appearance. Both have a bald red head (like a turkey).

The reason it is bald is that when a turkey vulture is feeding, it may have to stick its head into the carcass to find meat, and if there were feathers on its head, they would catch bits of the animal, including any parasites or bacteria it might harbor.

Black bear

Garter snakes may have 40 of them, opossums usually have about eight, catbirds often have four, and bats and porcupines have just one.

What are they? They're offspring. And different species of animals may have very different numbers of young when they become parents.

There are other differences as well in the reproductive habits of animals. Birds, butterflies and frogs lay eggs, but mammals, such as foxes, chipmunks and bears, have live young. Turtles may never meet their mothers, deer fawns may stay with their mothers nearly a year, and crows may stay together as a family composed of several generations.

But like most things that animals do, there is a purpose and a plan to the way they raise their families.

Frequently in their development, species face choices. To best ensure that a species survives, does it have a lot of young or just a few? Does it spend a lot of time taking care of the offspring or very little?

For instance, if a mother has lots of young each season, the chances are greater that some will survive to adulthood so they too can have families and keep the species alive. However, if she does have lots of young, she will have a harder time feeding them than if she has just a few babies. So which approach is best for species survival?

Animals do not consciously choose to use one approach or another. Essentially, life makes the decision for them. Let's say there is a species of animal in which litter size varies greatly. Some mothers have the biological trait – the genes – for having small litters, and some have the genes for large litters. But let's say that because of a permanent change in the availability of food, the young from small litters are more likely to survive. These young would inherit

Deer fawn

Red fox kit

the genes for small litters from their parents, and when they get to be parents, they will have small litters too. In not too many generations, most families of this species may be having small litters, and large litters might disappear completely.

If a species has been around a long time, chances are its method of reproduction has worked well. It is probably right for its situation.

For instance, birds need to be light to fly, and they need to eat a lot of food in comparison to their weight in order to have the energy to fly. They might have had a hard time surviving if they did not lay eggs. A mother bird might weigh too much to fly and feed if she were carrying her young inside her as they developed. She might also be more vulnerable to predators. So eggs that develop and hatch outside a mother's body are a good solution for birds.

Sometimes different approaches to reproduction work just as well. Some snakes, such as garter snakes, have live young, while others, such as milk snakes, lay eggs. The fact that snakes using each method are common seems to indicate neither method has a clear advantage for snakes.

Spring is the time for most offspring. Baby red foxes, woodchucks and river otters might be born just as spring starts. Mallards often lay eggs in April that hatch in May. Bobcats, skunks and chipmunks might have their young in May. Most songbirds, such as northern cardinals and tree sparrows, lay their eggs in May and June. Deer and little brown bats might be born in June.

For animals that are cared for by their parents, how long they remain in the nest or den varies by species. But as a general rule, the larger an animal is, the longer it will stay in the care of its parents.

An American goldfinch feeding its young

Bobcat kitten

The birds and the bees: Family planning in the animal kingdom

PAINTED TURTLE – From three to 14 eggs are laid in a hole dug on land in late spring, then the mother leaves the nest permanently. The eggs take about 75 days to hatch, then the young find water on their own.

GREEN FROG – As many as 4,000 eggs are laid in jelly-like masses in water, typically in mid-spring. The eggs hatch in three to six days, and the young may spend a year as tadpoles before changing into frogs.

HONEY BEE – The queen bee may lay up to 2,000 eggs a day and 200,000 eggs a season. An egg develops in a wax cell in the hive's brood area, hatching in about three days.

RED-TAILED HAWK – Typically, two or three eggs are laid in a stick nest built high in a tall tree. The eggs hatch in about a month. The young take first flights about 45 days later.

MEADOW VOLE – Also called field mice, these small mammals may have as many as 10 litters of live young a year, and there may be up to nine young per litter. The young may spend only two weeks being fed by their mother before they are on their own.

GARDEN SPIDER – Hundreds of eggs are laid in a sac on a leaf, twig or a support in the fall. The mother dies when the weather turns cold and the spiderlings overwinter in the egg sac.

WHITE-TAILED DEER – Usually two fawns are born in late May or early June. They feed on their mother's milk for about three weeks, then they begin to browse on vegetation. Males may leave their mothers by the fall. Females may stay with them up to two years.

RACCOON – Typically, three to seven young are born in a hollow tree in April or May. The young, which have the characteristic mask of raccoons soon after birth, venture from the den in about seven weeks to run and climb.

PUMPKINSEED – Using his fins, the male carves out a saucer-shaped nest in sand or gravel in shallow water in May or June. The female lays 1,700 or more eggs that hatch in as little as three days.

GARTER SNAKE – Females give birth to live young. The litter, delivered between July and September, may number 14 to 40 young. The newborns are immediately on their own.

Baby skunk

Magnolia warblers in their nest

What do you do if you find a baby animal?

You're walking through a field and come across a baby fawn standing motionless and alone in the tall grass. You find a baby bird beneath a tree in your back yard and there's no sign of either parent nearby. You accidentally uncover a nest of baby rabbits in thick grass on the edge of a lawn. What do you do?

Wildlife officials like to say, "If you care, leave them there."

In the majority of cases, an animal that appears to be orphaned really isn't. The mother may have wandered off in search of food, or she may be close by, waiting for you to leave. However, it may be that a young animal is out on its own. In the first days that animals leave the nest or den, they learn valuable lessons about finding food and protecting themselves that they will never have the chance to learn if you take them in without a good reason.

If the baby animal is truly abandoned or if you see an animal, baby or adult, that is obviously injured, call a veterinarian, an animal control officer or a licensed wildlife rehabilitator. If it's in your back yard, keep pet cats and dogs away while you wait.

If you decide it is necessary and safe to handle the animal, wear thick gloves and put it in a box with air holes and material, such as leaves or grass, on the bottom so that the animal does not slide around. Handle it as little as possible and call a wildlife officer.

However, it's not true that a mother animal will reject a baby if it has been touched by a human. She will take it back.

Most important, if the animal seems agitated or aggressive, do not handle it. Mammals, such as raccoons, foxes, skunks and bats, can carry rabies, a dangerous disease that can be passed to humans through an infected animal's saliva.

Eastern milksnake

Snakes can awaken primitive fears in us. The sudden discovery of a snake sunning itself on the front walk or slithering through the garden is enough to startle almost anyone.

But despite the impressive size of some native snakes of Pennsylvania (the eastern ratsnake can reach a length of up to eight feet), the chance of encountering one that could truly do you harm is very small. Of the 21 species of snakes that are native to the state, only three are venomous – the timber rattlesnake, the northern copperhead and the eastern massasauga. They are usually found in remote spots in the state, places humans rarely inhabit, such as high on rocky mountainsides.

Snakes live throughout the state. You're most likely to see a snake in the spring, when it is out of hibernation and basking in the sun. And the snake you're most likely to see is the garter snake, the most common snake in the state. Many snakes, such as the black racer, lay eggs, but others, such as the garter snake, have live young.

The northern water snake is the next most common snake in the state. A harmless snake, it is frequently mistaken for the poisonous water moccasin, also called the cottonmouth. However, there are no water moccasins north of Virginia.

Northern copperhead

The damage from being bitten by a non-venomous snake is more psychological than physical. The wound is typically no more serious than being scratched by a thorn of a rosebush.

However, each year in the United States on average, 8,000 people are bitten by one of the nation's 20 native venomous snake species. Only about a dozen of them die, though. Doctors who treat the victims say most of the incidents involve a copperhead.

Pennsylvania's most abundant venomous snake, the northern copperhead prefers rocky landcsapes in remote regions. It is generally not aggressive unless provoked. it is found throughout the state except in the northern counties.

The timber rattlesnake is the largest of the three venomous snakes native to the state, typically reaching three to four feet in length. It hunts mainly at night and is most active on nights with a full moon when it will wait in ambush for prey like chipmunks, squirrels or rabbits. It will then lunge at the animal, dig in its fangs, inject its venom, then withdraw to wait for the animal to die, which usually takes just a few minutes. If the animal is able to wander off, the rattlesnake will track it down by scent. It will then try to swallow the dead animal whole. Copperheads and timber rattlesnakes may eat only a dozen meals a year.

The smallest of the state's three venomous snakes, the eastern massasauga is only found in the westernmost

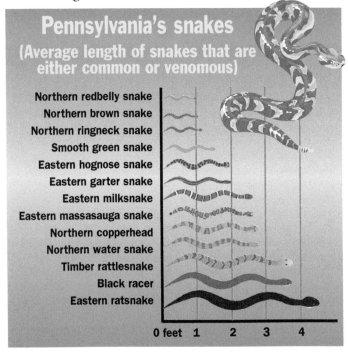

Garter snake

Pennsylvania's snakes
(Average length of snakes that are either common or venomous)

Northern redbelly snake
Northern brown snake
Northern ringneck snake
Smooth green snake
Eastern hognose snake
Eastern garter snake
Eastern milksnake
Eastern massasauga snake
Northern copperhead
Northern water snake
Timber rattlesnake
Black racer
Eastern ratsnake

0 feet   1   2   3   4

Smooth green snake

counties of Pennsylvania and it is one of the state's most endangered snakes.

Smaller snakes usually search for insects, slugs, earthworms, frogs or toads for their meals. Larger snakes may also hunt small rodents and birds.

All snakes swallow their prey whole, usually seizing the animal first with their jaws. However, a few snakes, including the black rat snake and milksnake, may wrap their body around the prey and squeeze it to make the kill.

Eastern ratsnake

People often mistake this for the rattle of a timber rattlesnake. But the rattlesnake has a very distinct rattle on it that when shaken sounds almost like a baby's rattle.

In winter, snakes in the Northeast hibernate in burrows or hollows or even in holes in the foundations of homes.

Most snakes are hatched or born in the summer, and they are left on their own to find food from the start. As snakes grow, they periodically shed their outer skin – like throwing away an old set of clothes

Some snakes, such as the milksnake and black racer, will vibrate their tails as a display to scare away predators.

– because they outgrow it. This is called shedding or molting.

Northern ringneck snake

Timber rattlesnake

Box turtle, 4 to 8.5 in.

Snapping turtle, 8 to 19 in.

Nature does not fix what isn't broken. Despite being slower and more awkward than almost any other animal its size or larger, turtles have changed remarkably little since they first appeared nearly 220 million years ago, well before most dinosaurs. Their basic design – a soft body contained inside a hard shell – has worked very well for them, providing them excellent protection against predators. In fact, some turtles may live 100 years or more.

The chief threat to turtles is humans. Many turtles are disappearing from the wild because they are collected for the pet trade. Some species are vanishing because their nesting sites, the places where they go to lay eggs, have been developed for businesses and homes.

Like snakes, turtles are reptiles, which means they cannot warm their bodies from the inside the way humans and other mammals do. Instead, reptiles must take action to warm up. For instance, they can bask in the sun. On cool spring days or on summer mornings, you will often see turtles in ponds doing just this, as they sit on top of logs or rocks.

Turtles have no teeth. They use their hard bills to scissor apart their food. In water, they eat aquatic insects, fish, frogs and plants. On land, earthworms, snails, grasshoppers, fruits and berries are all part of their diet.

A turtle's shell is both its mobile home and its protection. The top part of the shell, called the carapace, is attached to the turtle's backbone. The lower half of the

Turtles all lay eggs in the ground. Even a turtle that spends nearly all its life in water comes onto land to bury up to several dozen eggs in a shallow hole it digs in late spring or early summer. Most turtle eggs hatch from August to October, with a peak in September.

shell is called the plastron. Despite what you see in cartoons, turtles cannot leave their shells. The box turtle has a hinge on its plastron so that when it pulls in its head and legs to escape a predator the shell closes up, leaving almost no flesh visible.

Turtles have lungs but no gills. Some are able to breathe underwater by absorbing oxygen from the water through their exposed skin. In winter, many turtles settle down into the mud at the bottom of ponds or rivers to hibernate. Land turtles, such as the box turtle, dig beneath soft dirt or decaying leaves to hibernate.

Worldwide, there are nearly 320 species of turtles, including 13 that are native to Pennsylvania. The most common turtles found in the state, and the only two that can be found throughout the state, are the painted and snapping turtles. Snapping turtles are the largest of the Pennsylvania turtles. In 2006, one was found in Wayne County that weighed more than 60 pounds.

The painted turtle lives almost anywhere there are ponds, lakes or slow-moving streams or rivers. Snapping turtles, which spend most of their time underwater, are rarely seen except during the breeding season when they come onto land to lay eggs.

Some Pennsylvania turtles are found in only a few locations in the state. For instance, the bog turtle, eastern mud turtle and eastern redbelly slider are found mainly in the southeastern counties. The Blanding's turtle has only been found in one county, Erie County, in the northwest corner of the state.

Eastern musk turtle, 3 to 5.5 in.

Spotted turtle, 3.5 to 5 in.

Wood turtle, 5 to 9 in.

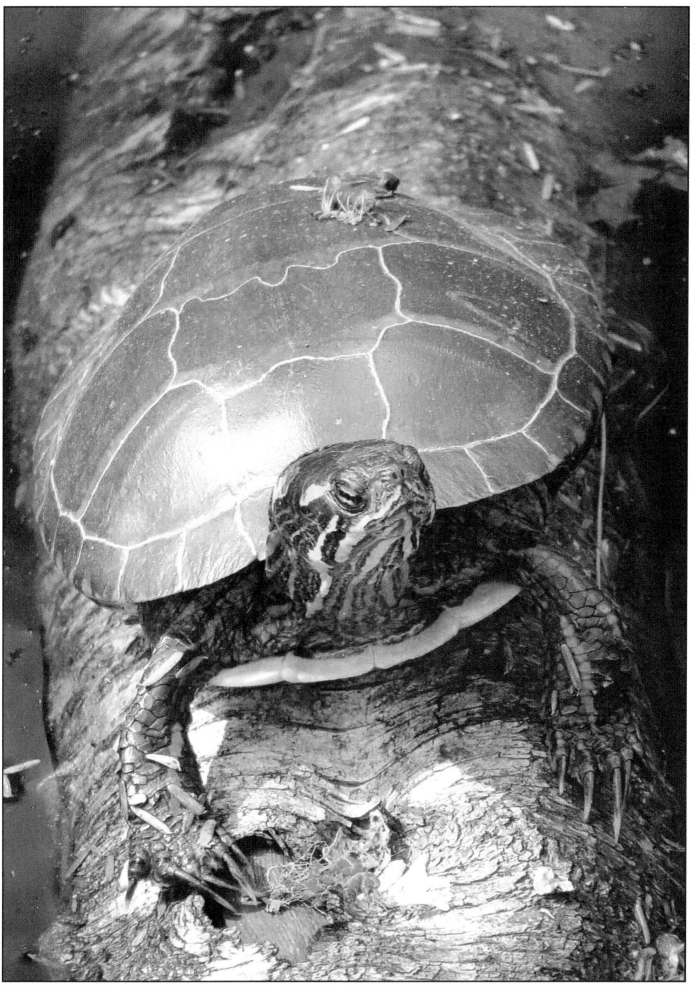

Eastern painted turtle, typically 4 to 10 in.

If there were an animal Olympics, frogs would certainly be competitors in the standing long jump. Some can leap more than 30 times their body length.

Lots of animals are specialists physically. They have a feature that gives them an advantage over other animals in finding food or defending themselves, and that's why their species has survived over the centuries. Frogs use their tremendous jumping ability both to escape predators and to pounce on a meal.

Frogs are amphibians, which means they spend part of their lives in water and part on land. They are also among the animals that undergo metamorphosis, which means a "change in form." As young tadpoles (also called polliwogs), they live in water and look like small fish with large rounded heads. They have gills, tails and no legs. But eventually they develop lungs and legs, they will lose their tails and, as frogs, they will be able to live on land.

There may be nearly 4,800 species of frogs worldwide. (This includes toads, which technically are a kind of frog.) There may be 17 species of frogs found in Pennsylvania.

Most adult frogs have large hind legs for jumping, hind toes that are connected by webbing, no claws and tiny teeth (if they have any at all).

Some, such as the northern cricket frog and the southern leopard frog, are rare enough in the state that you will probably

Bullfrog, 3 to 8 in.

never come across one in your lifetime. But others, like bullfrogs and green frogs, are so common that you can't miss them. In spring or summer, walk down to the edge of any pond and you may see a green frog sitting very still on the bank or a bullfrog peering at you from the water, its two bulging eyes and a bit of its head breaking the pond's surface.

Spend any time around a pond and you'll also hear frogs calling, or chorusing, especially during breeding season, which usually peaks in the spring. Male frogs are the ones making all the noise. They call to attract females and to announce that a territory is theirs. A frog enhances its call by using a loose pouch of skin at its throat that it can fill with air. When it calls, the sound enters this air-filled chamber and reverberates, like an echo in a cave, becoming even more intense. A male frog might repeat its call thousands of times in one night.

Nearly all female frogs lay jellylike eggs in water, often thousands of eggs at a time, which eventually hatch to become tadpoles.

In winter, frogs become inactive, settling down into the mud at the bottom of ponds or taking refuge under piles of dead leaves or in underground tunnels on land. In the coldest weather, some even partially freeze but are still able to thaw out and resume their lives in the spring.

Most frogs eat insects, such as ants and flies, as well as worms and snails. However,

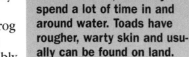

Frog or toad?

You can usually tell the two apart by the texture of their skins. Frogs tend to have smoother skin, and they spend a lot of time in and around water. Toads have rougher, warty skin and usually can be found on land.

Leopard frog, 2 to 5 in.

American toad, 2 to 4.5 in.

Green frog, 2 to 4 in.

the bullfrog, which is the largest native Pennsylvania frog, will eat fish, snakes, mice and even small birds or newly hatched turtles that venture too close.

In recent years, biologists have noticed a growing problem among amphibians. Many have deformities, such as missing or extra limbs. In addition, populations of frogs in some places in the world are declining dramatically. Researchers are not sure what the reason is for the troubling problem. A leading theory is that it may be due to a disease or pesticides.

Gray treefrog, 1.5 to 2 in.

Humans vs. animals

The peak performance of humans as of 2015 compared with the observed performance of animals in the wild

Marathon
(26 miles, 385 yards)
Human: 2 hr., 2 min., 57 sec.
Pronghorn antelope: About 45 min.

High jump
Human: 8 ft., .5 in.
Killer whale: 15 to 17 ft.

Speed on land
(100 meter dash)
Human: 9.58 sec.
Cheetah: About 3.2 sec.

Speed in water
(100 meter freestyle)
Human: 46.91 sec.
Sailfish: About 3.3 sec.

Long jump
Human: 29 ft., 4.3 in.
Snow leopard: About 50 ft.

Vernal pools

On the first warm night that follows a heavy rain each spring, an amazing event occurs. Certain frogs and salamanders that are rarely seen at any other time of the year come out of hiding and begin moving through the woods in great numbers.

Crowds of frogs. Parades of salamanders.

They are migrating to pools of water left by melted snow on forest floors. They may cross roads, hike up hills and climb over rocks and fallen trees in their determination to reach these temporary patches of water, called vernal pools.

Because the water dries up in the heat of summer, these pools

Spotted salamander 6 to 10 in.

do not contain fish. That makes them ideal breeding ponds for these amphibians, since there will be no predatory fish in the water to eat their eggs or their developing young.

In Pennsylvania, wood frogs make this trek, as do spotted, blue-spotted, marbled and Jefferson salamanders.

During March and April, you can locate some of these breeding pools by the sounds that come from them at night. Wood frogs will chorus, making a noise that resembles ducks quacking.

Once these amphibians reach their vernal pool, which they may return to year after year, they must court, breed and lay their eggs in a short period of time.

They are active by day, but most of the activity is at night, especially among the salamanders.

If you can locate a vernal pool in the spring, take a flashlight, shine it on the water after dark, and you might see dozens of these creatures swimming about.

Wood frog, 1.5 to 3.5 in.

What monarch butterflies do each fall is just about the definition of impossible.

Each September, tens of millions of these regal orange and black butterflies leave their summer breeding grounds in the United States and Canada and flutter off toward their winter home on a few remote mountainsides in Central Mexico. They return to the very same trees that previous generations returned to in their southern migrations for perhaps thousands of years. Then, in March, they may lift off nearly all at once, forming great clouds of butterflies that fill the sky as they return north.

But the astounding thing is that adult monarchs may live less than one month in summer. That means that none of the monarchs that leave Mexico in the spring are among those that return in the fall. Instead, it is often their great-great-grandchildren that will somehow find their way back to those few Mexican mountainsides with no one in their band of migrating monarchs ever having been there.

How do they do it? Is it in their genes? Are they guided by the Earth's magnetic field? Or is it just one of the mysteries of animal instincts?

Butterflies, like most insects, go through a metamorphosis, or a change in appearance, except that for butterflies it is one of the most striking changes.

Butterflies have four stages to their lives – egg, larva (also called a caterpillar), pupa (also called a chrysalis), then adult butterfly.

In the pupa stage, the caterpillar forms a shell-like covering around itself and hangs by thin threads from a twig or leaf. Inside, the

A monarch butterfly rests in a field during migration

caterpillar can change from a fairly clumsy, slow-moving and perhaps, to some, even ugly creature into what can be a beautiful, brilliantly colored butterfly, carrying the colors of a rainbow on its fragile wings.

Butterflies feed on nectar, a sweet liquid produced deep inside flowers. To do this, a butterfly uses its proboscis, a long narrow tube like a drinking straw that can be almost as long as its body. When it's not in use, the proboscis is kept curled up where you would expect the butterfly's mouth to be.

Butterflies need the heat of the sun to warm their bodies, so they tend to be most active during the middle of the day. At night, they rest.

Butterflies have three body parts – the head, the thorax or middle section, and the abdomen. They have six legs, and four wings – a pair each of front and back wings.

The colors on their wings are created by tiny colored scales that fit together like tiles in a mosaic.

Most adult butterflies live only two to four weeks, surviving long enough to mate and produce eggs. The female searches for certain kinds of plants on which to lay her eggs, since they are food plants for the young caterpillars. Monarchs choose milkweed. White admirals like wild cherry and poplar leaves. Tiger swallowtails prefer black cherry leaves.

There may be 20,000 species of butterflies in the world. In Pennsylvania, there are about 155 species that have been seen, including 25 to 30 species that are only occasional visitors to the state, carried here by a large storm or on strong winds from other states or regions of the country.

Butterfly or moth?

There are lots of differences. Butterflies usually fly during the day. Moths usually fly at night.

Butterflies tend to rest with their wings folded up over their heads or stretched flat to each side. Moths often lay their wings flat and behind them while resting.

The antennae of butterflies are thin and end in a knob. Those of moths do not end in a knob and are often feathery. Butterflies usually have bolder and brighter colors than moths, although there are some strikingly colored moths.

Moth

Butterfly

Baltimore checkerspot

Eastern black swallowtail

The science of butterfly wings

Butterflies are the super-models of the insect kingdom, garbed in spectacular colors and sensational designs.

However, all that artistry has more purpose than to please the human eye. The hues, patterns and other features carried on those fragile wings have multiple purposes, all of them critical to survival – discouraging predators, generating heat and wooing the opposite sex.

For instance, some butterflies have evolved to have a feature on their wings called eyespots, which help to discourage predators. The little wood satyr has a pair of eyespots on the edges of both its front and back wings. They are spots that may appear to a poor-sighted predator to look like eyes. They can startle a predator like a bird or snake, buying the butterfly an instant of time in which to make its escape.

Swallowtail butterflies have small projections at the ends of their wings that look like tails (thus their name). These features evolved to attract the attention of predators who may attack the false tails, which tear off easily, giving the butterfly a chance to escape.

Butterflies, like snakes, frogs and turtles, don't have a way to create heat inside their body, the way humans do. They are the temperature of their surroundings and often have to warm up by basking in the sun.

People know that if they wear dark colors on a sunny day, they are likely to be warmer than if they wear light colors. Dark colors, such as brown and black, absorb more sunlight than light colors, such as yellow or white, and the absorbed sunlight creates heat.

The tail of the eastern tiger swallowtail, which tears away easily, can be a misleading target for predators.

The dark coloring of a mourning cloak absorbs sunlight, which creates heat in cold weather.

The eyespots of a little wood satyr may fool predators into attacking the outer wing edges instead of the head.

For butterflies, which need a way to generate heat on a cool day, this trait is important. Nearest to their bodies, most butterflies have dark colors. That's because butterflies need a body temperature of at least 80 degrees to fly well. Their muscles won't work properly otherwise. For this reason, butterflies that fly in the early spring or late fall or that overwinter as adults in the state, such as mourning cloaks, tend to have darker coloring.

However, the need for dark colors farther out on the wings isn't as great as it is near the body. That's because very little blood circulates in the wings and they are only able to transfer heat for a short distance.

Reproduction does require bright colors and distinctive patterns somewhere on a butterfly, not only to attract the eye of the opposite sex, but also to identify it as being of the same species. Females choose males based on how nicely and brightly colored they are. Males make the same kind of choices.

Once you know all this, an eastern tiger swallowtail's features makes a lot more sense: dark near its body for heat, false tails on its wings to foil predators, and fantastic yellow coloring with black patterns on most of its wings to lure a mate.

The design of butterflies, besides being functional, is quite ancient. Human beings have only been around about two million years, but physically and intellectually we are a species that has changed dramatically. Butterflies, on the other hand, have been flitting from flower to flower and sporting colorful patterns on their wings for more than 40 million years, with few evolutionary alterations. That must mean their design has worked well for them.

CHECKLIST
Common butterflies

Mourning cloak
3 – 3.5 in.
wingspan
☐

Monarch
3.5 – 4 in.
☐

Eastern tiger swallowtail
3 – 5.5 in.
☐

Spicebush swallowtail
3.5 – 4.5 in.
☐

Eastern black swallowtail
2.5 – 3.5 in.
☐

Red-spotted purple
3 – 4 in.
☐

Red admiral
2 – 2.5 in.
☐

Great spangled fritillary
2 – 3 in.
☐

Baltimore checkerspot
1.5 – 2.5 in.
☐

Pearl crescent
1 – 1.5 in.
☐

Eastern comma
1.5 – 2 in.
☐

American copper
About 1 in.
☐

Painted lady
2 – 2.5 in.
☐

Spring azure
About 1 in.
☐

Cabbage white
1.5 – 2 in.
☐

Clouded sulphur
1.5 – 2 in.
☐

Silver-spotted skipper
1.5 – 2.5 in.
☐

Common ringlet
1 – 2 in.
☐

Little wood satyr
1.5 – 2 in.
☐

Banded hairstreak
1 – 1.5 in.
☐

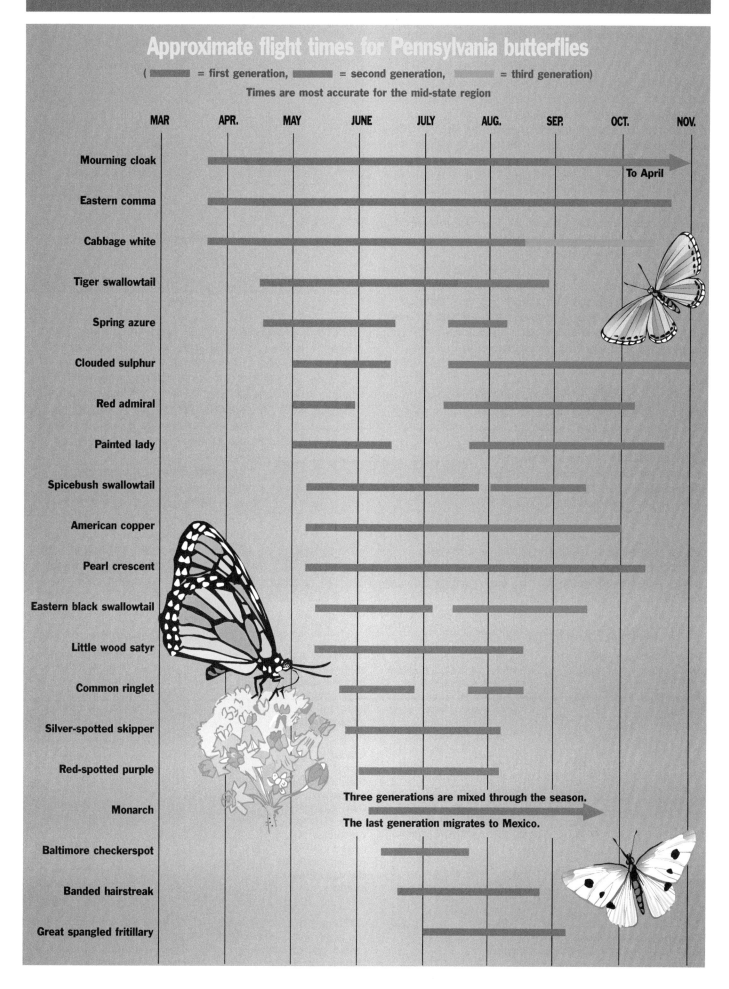

Approximate flight times for Pennsylvania butterflies

(▬▬▬ = first generation, ▬▬▬ = second generation, ▬▬▬ = third generation)

Times are most accurate for the mid-state region

	MAR	APR.	MAY	JUNE	JULY	AUG.	SEP.	OCT.	NOV.

Mourning cloak — To April

Eastern comma

Cabbage white

Tiger swallowtail

Spring azure

Clouded sulphur

Red admiral

Painted lady

Spicebush swallowtail

American copper

Pearl crescent

Eastern black swallowtail

Little wood satyr

Common ringlet

Silver-spotted skipper

Red-spotted purple

Monarch — Three generations are mixed through the season. The last generation migrates to Mexico.

Baltimore checkerspot

Banded hairstreak

Great spangled fritillary

At most family reunions, get all the relatives together in one room and you can usually see a resemblance.

But put a caterpillar and a butterfly side by side, or a tadpole and a frog next to each other, and not only is there no family resemblance, there's not even a species resemblance. Why does this happen?

This dramatic change in appearance, this metamorphosis, evolved in nature as a way of ensuring that these species would survive.

Many features of plants and animals – the long legs of a great blue heron, the rich color of a rose, the sharp beak of a hawk – evolved or developed because they give a species a better chance for survival. They are features that give it an edge in finding food or in reproducing or in protecting itself from predators.

Metamorphosis is no different. Species that change form (and all amphibians and nearly all insects go through some type of metamorphosis) can occupy two very different places in the ecosystem. The young may eat one kind of food, go through metamorphosis, and then eat a different kind of food as adults. That means there is more food available for the entire species, increasing its chance for survival.

Tadpoles are vegetarians, but adult bullfrogs eat almost anything, from plants to flying insects. Most caterpillars feed on leaves, but adult butterflies sip the rich nectar they find in flowers.

Having different forms at different stages of life can help in other ways as well. Caterpillars do almost nothing but eat and grow, and they are well-designed to do just that. They are able to avoid predators to some extent because most can blend into vegetation. They are also able to blend into their environment as pupae, the stage in which they are surrounded by mummylike coverings and

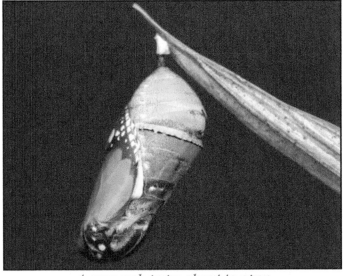

A monarch pupa about to open

undergo the changes that make them adult butterflies. Then, in the butterfly stage, they can fly to new areas to lay eggs and spread their species.

The changes tadpoles and caterpillars go through during metamorphosis are striking. Bullfrog tadpoles, which will be going from a life spent entirely in water to a life on land and in water, have to develop limbs and lungs during the change, and their tails have to be reabsorbed into their bodies.

Caterpillars, during the pupa stage, have to lose their many legs as well as develop wings and sex organs. They will also develop new mouth parts to draw nectar from flowers as adult butterflies.

For frogs, the metamorphosis usually lasts from several days to a few weeks. For butterflies, it may last from a few days to several months.

Monarch caterpillar

Monarch metamorphosis

In about 10 days, the adult butterfly emerges from the pupa, unfolds its wings, which slowly harden, then it flies away.

In the pupa stage, wings develop and the mouth parts change from chewing parts to sucking parts so that the adult butterfly can sip flower nectar.

The caterpillar attaches itself by a silk thread and tiny hooks on its abdomen to a plant stalk and then sheds its skin one last time to reveal the pupa beneath.

Eggs are laid by female monarchs on the undersides of the leaves of milkweed plants. Each larva, or caterpillar, will eat its way out of its egg within a week and then begin to feed on the milkweed.

For the next two to three weeks, the caterpillar will feed and grow, shedding its skin every so often, a process called molting.

Bullfrog metamorphosis

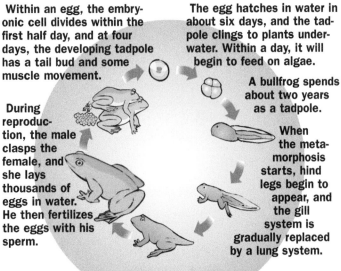

Within an egg, the embryonic cell divides within the first half day, and at four days, the developing tadpole has a tail bud and some muscle movement.

During reproduction, the male clasps the female, and she lays thousands of eggs in water. He then fertilizes the eggs with his sperm.

The frog is a mature adult in about three years. Adult bullfrogs typically weigh about one pound.

The egg hatches in water in about six days, and the tadpole clings to plants underwater. Within a day, it will begin to feed on algae.

A bullfrog spends about two years as a tadpole.

When the metamorphosis starts, hind legs begin to appear, and the gill system is gradually replaced by a lung system.

Within two weeks, the metamorphosis is usually complete. With its lungs, a frog can breathe air.

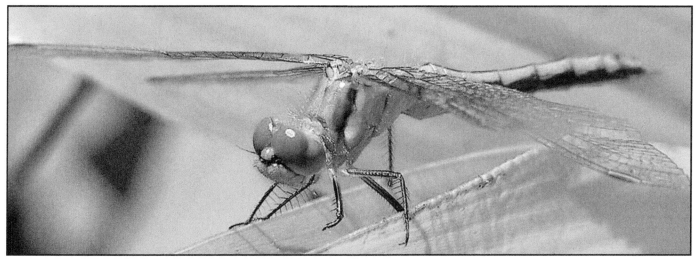

Ruby meadowhawk, female

As if the passion of the public for boldly colored winged creatures can't be satisfied by birds and butterflies alone, dragonflies are now attracting their faithful followers.

And why not? They are elegant creatures, capable of mid-air acrobatics that few, if any, birds can match. They are also the friends of anyone who doesn't like mosquitoes. The reason dragonflies spend so much time hovering over ponds is that they are hunting mosquitoes, which are their primary food.

Twelve-spotted skimmer, male

Worldwide, there are about 6,000 species in the order Odonata, which includes dragonflies and damselflies. There may be 170 species native to Pennsylvania.

Like birds, dragonflies fly and lay eggs. Some, like the green darner, even migrate to warmer regions when the weather turns cold. However, like butterflies and other insects, dragonflies undergo metamorphosis, changing from waterbound larvae to flying adults. Unlike butterflies, though, there is no pupa stage.

The eggs of dragonflies, which are laid in or near water, hatch in five to 10 days. The larvae that emerge then begin the process of eating and growing. They consume everything from mosquito larvae to tiny fish, capturing their prey with a unique lower lip that has its own claw. The lip can shoot out to almost a third of a larva's body length to grab a meal.

Larvae that hatch in the summer often spend the winter growing before transforming into adult dragonflies the following spring.

When it is ready to make the change to a flying adult, the larva climbs out of the water, takes hold of a reed or twig, and sheds its skin, allowing the soft, folded wings beneath to slowly fill with blood and harden. Within an hour or two, the adult dragonfly takes to the air.

Many dragonflies live only about a month as adults, time enough to mate (usually accomplished in mid-air) and to lay eggs before dying.

The flight abilities of dragonflies are indeed impressive. The larger species, such as the darners, may be capable of flying up to 50 miles per hour. Seemingly tireless, they may spend most of the day in the air, catching prey on the wing.

What makes them such powerful aviators? For one thing, like a hummingbird, a high percentage of their body mass is dedicated to flight. A sexually mature male may be more than 60 percent flight muscle.

Dragonflies can take off backward, they can launch vertically like a helicopter, and they can hover motionless for more than a minute. They can also stop on a dime.

Dragonfly or damselfly?

Dragonflies and damselflies are both found flying over ponds, swamps and other still water bodies. You can tell the two apart by how they hold their wings when at rest. A dragonfly will stretch them out from side to side. Damselflies, which have thin, needle-like bodies, rest with their wings folded in back of them close to their sides. Contrary to myth, neither dragonflies nor damselflies bite humans.

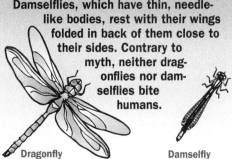

Dragonfly   Damselfly

Halloween pennant, male

Dragonflies mating

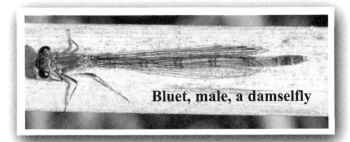

Bluet, male, a damselfly

Green darner, female

American lady

Monarch

Eastern tiger swallowtail

What tangled webs they weave. Worldwide, nearly 44,000 kinds of spiders have been identified, including about 3,000 in the United States. There may be 600 species native to Pennsylvania.

You can tell a spider by its two body sections and eight legs. (Since insects have only six legs, spiders are not true insects.)

Interestingly, most spiders also have eight eyes, with some found on the side instead of the front of their bodies. Yet, spiders have poor eyesight generally.

Many spiders are active only at night, and most are not very aggressive. Even those with the strongest venom will attack humans only as a defensive measure, preferring to retreat from battle whenever possible.

Because of the Northeast's chilly winters, most spiders found in Pennsylvania live just one season, hatching in the spring, breeding and laying eggs in the summer, and dying in the fall. But a certain number manage to live through the winter. They may try to escape harsh weather by seeking refuge in the cozy interior of your home or apartment. A large house can be home to hundreds of spiders in winter, with the human residents barely aware of their presence.

Most spiders you are likely to encounter in Pennsylvania would be too small to pose a threat. As a general rule, only those with a body over half an inch in length (not including legs) have fangs capable of penetrating human skin. But that does not mean they are dangerous. Even larger spiders that bite are not usually a serious danger.

Just about all spiders are venomous. Even the smaller ones use venom to weaken or kill their prey. But only a

The multiple eyes of a spider

few spiders have strong enough venom to cause harm to a human being.

The two most venomous spiders to be found in the United States are the brown recluse and the black widow. However, most people recover if bitten by either.

The brown recluse, which has a violin-shaped mark on its back, is only occasionally seen in Pennsylvania. It sometimes arrives in the state as a hitchhiker on furniture or clothes brought from areas where it is more common, such as the Midwest. Its bite can have serious effects and

Among the largest spiders you might cross paths with in Pennsylvania are these two, shown their actual size. They can be found in homes or sheds near forested areas. Their venom is not poisonous to humans, but their bites can leave a sore.

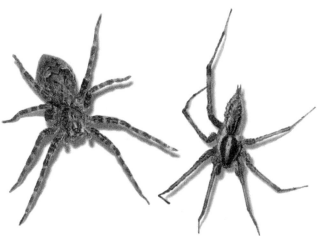

Brownish-gray fishing spider, female, 1 in.

Wolf spider, female, 1 in.

Nursery web spider, female, .6 in.

must be treated quickly.

Two species of black widow spiders are found in the state, the northern and more common southern varieties. Both are black with a red mark on their abdomens that is shaped like an hourglass. In the northern, the hourglass is divided at its narrowest point.

The female black widow earned her name because soon after mating she may eat the male. The males of

Crab spider, female, .4 in.

most species of spiders are not treated so badly.

One of the largest spiders someone in Pennsylvania` might cross paths with indoors is the wolf spider, often a resident of garden sheds and suburban gardens, but sometimes an invader of homes. Its body can be more than an inch long. While painful, its bite usually does not have long-lasting effects. In most species, the female is larger than the male.

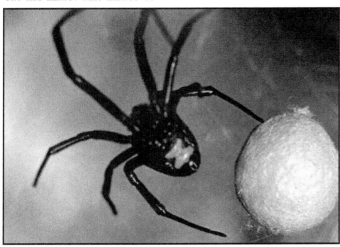
Southern black widow spider, female, .5 in.

Brown recluse spider, female, .4 in.

Spider webs

The delicate webs that decorate dew-covered meadows or that float weightlessly in the corners of attic ceilings are actually wonders of engineering. Pound for pound, the silk in a spider web is stronger than steel.

Spiders produce the silk in their abdomens. It starts out as a sticky gel made of pure protein that is created in an array of glands. The gel is pushed down tubes that lead to ducts called spinnerets that will shoot it out of the body.

By contracting its abdomen muscles, a spider is able to extrude the gel from thousands of tiny spigots on the spinnerets. In the process, the gel changes to a solid. And like the many fibers that come together to create hemp rope, so the many strands exiting the spigots come together to create the finished silk line.

The final strand of silk may be less than a tenth the thickness of a human hair, but it can hold an astounding amount of weight.

As a comparison, a rope that is an inch thick would have to support as many as 15 automobiles to have the same strength as some spider silk.

Aside from creating webs with their silk to snare passing flies and other prey, spiders use it as a safety line so that they can drop down from ceilings or branches. They also use silk to wrap their eggs for protection and to wrap prey for a later meal. A spider may produce up to eight kinds of silk, each for a different use.

Watch a colony of ants around their nest in a field or a swarm of gnats on a pond in summer and almost inevitably the question occurs to you.

Just how many insects are there in the world?

Well, science has an answer – or at least a rough estimate. At any given moment on the planet, there are about 10 quintillion of them. That's roughly 10,000,000,000,000,000,000 mayflies, millipedes, monarchs, mantises and more.

An impressive number, it's the best guess of famed Harvard entomologist E. O. Wilson. To put it in perspective, 10 quintillion is more than the number of seconds that would have ticked on a clock if it began keeping time the moment the universe began.

Worldwide, nearly a million species of insects have been identified. That's more than the number of all other species of animals combined. (There may be nearly 5,500 species of mammals and about 10,000 species of birds.) However, scientists say they have recorded only a fraction of all the insect species that actually exist, which they say may number more than two million.

Insects are found in just about every environment on Earth, including in Antarctica where several dozen species of insects have been seen.

If for no other reason than their sheer numbers, insects are arguably

Bumblebee, .7 in.   *Honey bee, .5 in.*

Bee or wasp?

Both can sting if bothered, but bees tend to have hairy bodies while wasps usually do not. Also, many bees, such as bumble bees and honey bees, have on their hind legs "pollen baskets" made of stiff hairs, where they carry the pollen they collect on their visits to flowers. Wasps do not have them. Hornets and yellow jackets are kinds of wasps.

Bumblebee   **Paper wasp**

the most successful animals on the planet. Sure, they can be some of the most repellent creatures known to humankind, but they are perhaps more necessary to our survival than any other group of animals. They play a critical role in the ecology of Earth in ways that most people don't realize. Should insects all disappear at once, humans would soon follow since the lives of insects and humans are so intertwined.

Why? Insects are eaten by small animals, which are in turn eaten by bigger animals, which might ultimately be eaten by humans, so the entire food chain would fall apart. Also, insects pollinate the majority of fruits and vegetables humans eat, including apples, oranges, blueberries, cranberries, cherries, melons and almonds. Insects also break down dead plants and animals as well as animal wastes into the nutrients that fertilize the soil, allowing things to grow.

Then, there is the less measurable spiritual loss the disappearance of beetles and butterflies, ladybugs and waterbugs, would create. Most flowering plants, including field wildflowers, are pollinated by insects, so they would also disappear. Indeed, some of the poetry would go out of our daily lives.

Insects are the ultimate survivors, despite their small size and short lives. (Some adult mosquitoes are lucky to live two weeks). In fact, their brief lives

Woolly bear caterpillar, 2 in.

Praying mantis, 3.5 in.

True katydid, 2 in.

help make them so successful in surviving change. Insects reproduce quickly and in great numbers. So if there is any change in their surroundings – for instance, if the temperature warms dramatically – the chances are good that they will produce some young from the thousands of eggs they might lay (some termite queens produce 30,000 eggs a day) that have the right combination of genes to overcome the warmer weather. Some of those young will survive, reproduce more young with similar genes, and soon the local population will have adapted to the warmer climate and be thriving again.

Pennsylvania leather-wing

Species that take a relatively long time to reproduce, such as some birds, which may have just two or three young a season, would not be able to adapt to dramatic changes in their habitat so quickly. Such changes might completely wipe out their local population.

All insects have certain things in common. They have six legs, three body sections and antennae. Spiders, with their eight legs, are not true insects. Most insects can hear, taste, touch, smell and see, but many of them do these things in very different ways than humans. Crickets hear with their knees. Flies taste with their feet. Some

moths and beetles smell with their antennae.

Most insects have two main eyes and two or three smaller, simpler eyes. The main eyes are sometimes made up of thousands of individual lenses, each of which produces an image. Human eyes have one lens each, so the human brain has to interpret only the two images that its eyes send it. But a dragonfly's eyes may have 28,000 lenses each, which means its tiny brain has to interpret that many separate signals in order to understand what it is seeing.

Nearly every insect goes through metamorphosis, a change in appearance marking a new stage of its life. Most insects look different as adults than as young emerging from eggs. Many are flightless when young but have wings as adults.

Big bugs

Here are a few of the larger insects that can be found in Pennsylvania. They are shown their actual size.

Fiery searcher

Eastern dobsonfly

Dogday harvestfly (cicada)

Luna moth

Fireflies

Also called lightning bugs, these beetles produce light in the tips of their abdomens by combining chemicals they produce and store in their bodies. Their blinking in summer is part of their mating process. Males fly about, flashing their lights to attract females that sit on leaves and branches.

There are nearly 200 species of fireflies in North America. A few don't flash, but those that do have their own pattern of flashes to help males and females of the same species identify each other.

In the East, the male of one of the most common fireflies (*Photinus pyralis*) flashes in low flight over grass about every six seconds, creating a yellowish "J" that lasts about half a second as he lifts into the air. The female, which sits on vegetation nearby, responds about two seconds later with a half-second flash of her own, an invitation for him to approach.

Ants

Like bees and other social insects, ants live in organized societies in which the members work together to keep the colony functioning. In all ant colonies, one queen ant is responsible for laying the eggs and producing the young.

Black carpenter ants, which are about a half-inch long, build their nests in dead wood, creating a series of tunnels and rooms. Unlike termites, they do not eat wood. They only tunnel into it.

Little black ants, which may be one-fourth the size of carpenter ants, usually build their nest underground, forming a little mound at the opening.

Mosquitoes

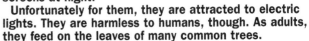

It is only the female mosquito that will bite you. The males feed mainly on flower nectar and fruit juices. The females need a meal of blood before laying eggs. Mosquitoes are most active at dawn, at dusk and at night. They hatch from eggs that develop in water, such as rainwater that collects in roadside ditches or in discarded tires.

Grasshoppers and crickets

The "songs" of grasshoppers, like those of crickets, are the music of summer nights. However, these insects don't sing the way humans do. Grasshoppers rub a rough part of a back leg against their wings to create their music, and crickets rub their wings together. They make these noises to attract females and to chase away other males.

Ladybugs

Probably the best known of beetles, ladybugs are among the most valuable beetles to humans. They generally do not eat plants, but they do eat small insects that eat plants, which makes them important to farmers. In this region, you might see ladybugs with as few as zero spots or with as many as 19.

June bugs

Also called May beetles, June bugs are often seen buzzing around porch lights, or they are heard crashing into windows and screens at night.

Unfortunately for them, they are attracted to electric lights. They are harmless to humans, though. As adults, they feed on the leaves of many common trees.

House flies

Unwelcome guests at a picnic and a noisy distraction when they get inside a home or apartment, house flies live short lives and reproduce quickly. But they do not bite.

Their eggs can hatch in just a day, and the larvae can grow to adulthood in a week. A house fly may be hatched, mature, reproduce and die in the space of just a month, which means several generations of house flies can emerge during one summer.

Honey bees

Honey bees live in ordered communities in which each bee has a job. A honey bee hive may have 50,000 members. There is only one queen bee, and her job is to lay eggs. There may be 1,000 male drones in the hive, and their only job is to mate with the queen, although only one will be successful. The rest of the hive's members are female worker bees, and they can have many jobs during their brief lives. Here is the typical life cycle of one worker bee in summer.

DAY 1–3
The egg is laid and develops within a wax cell.

DAY 4–20
The larva emerges from the egg, feeds, and then develops into an adult worker bee.

DAY 21–23
The worker bee's first job is as a cleaner, readying the brood cells for the next batch of eggs.

DAY 24–27
It becomes a nurse, feeding the older larvae.

DAY 28–34
It continues as a nurse, feeding the younger larvae and the queen.

DAY 35–41
It becomes a food collector, searching outside the hive for pollen, a rich source of protein.

DAY 42–death
It continues as a food collector, searching outside the hive for nectar, a rich source of energy.

At different times, the worker bee may also be a fanner (beating its wings to keep the hive cool), an undertaker (removing the bodies of dead bees from the hive), a soldier (guarding the hive's entrance) or a builder (creating wax to add to the hive walls).

Japanese beetle

Cecropia caterpillars

Periodical cicada

Asian ladybug

Red-legged grasshopper

Green bottle fly

Same planet, different worlds.

That's a good definition of ecosystems, the largely self-contained collections of plants, animals and ecological conditions that cover the landscape like different countries on a map.

A pond with bullfrogs, cattails and dragonflies can be near a forest where you might see hemlocks, wolf spiders and white-tailed deer. And just a stone's throw away can be a field where you might find monarch butterflies, oxeye daisies and meadow voles.

An ecosystem is a collection of living and non-living things, including plants, animals, ponds, streams, rocks and dirt, that all interact as a unit.

An ecosystem is created by a particular combination of water, land features, sunlight and temperature. Wherever you find that set of physical conditions in a region, you're likely to find the same general group of plants and animals.

An ecosystem can include a stream or a pond, or it can include a combination of a stream and a pond. An ecosystem can include a meadow in a valley or a meadow on a mountainside.

An ecosystem can be as small as a single small pool of water left from melted snow on a forest floor in spring, or it can be as large as the whole forest, including the pool of water. Ecosystems are also found in cities. A vacant lot between two buildings may eventually become home to a variety of wildflowers, insects, birds and small mammals.

Ecosystems do not have barriers between them, though. They overlap and share resources. Insects, birds and other animals move between ecosystems.

In nature, plants and animals depend on each other for food. Many birds eat insects. Many insects eat the leaves of trees and other plants. Trees and other plants draw nutrition through their roots from dead plants and animals that decay on the ground, and they produce nuts and seeds that birds and other animals eat. This is called a food web.

After a long period of time, a balance or a harmony in the food web in an ecosystem is often reached so that populations of the different species stay within certain limits. For instance, if a predator, such as foxes, grows too plentiful, the prey, such as deer mice, might die off. Without enough food, foxes might decline in number and the mice have a chance to grow more plentiful again, allowing foxes to grow in population once again. It is a continuing cycle.

Food web

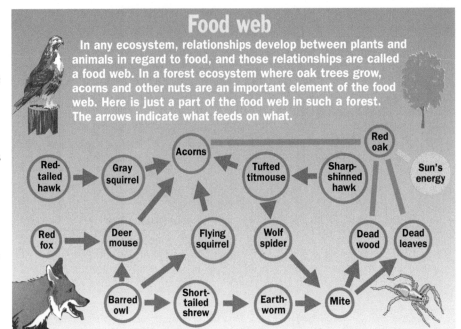

In any ecosystem, relationships develop between plants and animals in regard to food, and those relationships are called a food web. In a forest ecosystem where oak trees grow, acorns and other nuts are an important element of the food web. Here is just a part of the food web in such a forest. The arrows indicate what feeds on what.

Ecozones of Pennsylvania

In an ecozone, the same general physical conditions exist. Some ecosystems may be found in all ecozones, and one ecozone may contain dozens of different ecosystems.

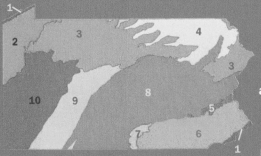

10 Allegheny plateau
Hilly and forested landscape with abundant coal deposits and many farms

9 Allegheny Mountains
High rugged plateau with steep-sided mountains and hills that create cooler and wetter conditions than in adjacent regions

8 Ridges and valleys
A series of narrow ridges and valleys with forests on the ridges and farms in the valleys

1 Great Lakes and Delaware River lowlands
Flat low plains with higher population density and urban development than much of the rest of the state

7 Blue Ridge Mountains
Narrow ridges and hilly plateaus with elevations to 2,200 feet; forests of hardwoods as well as spruce and fir

2 Erie plains
Rolling plains of farms mixed with forests of oak, maple beech and hemlock

3 North central Appalachians
Forested elevated plateau on erosion resistant sandstone

4 Northern Appalachian plateau and uplands
Farms, pastures, and forests among rolling hills and open valleys

5 Northeastern highlands
Low mountains with forests of maple, beech, birch and hemlock

6 Northern piedmont
Open valleys and irregular plains among low rounded hills

Examples of Pennsylvania ecosystems

Cattail marsh

Often found where soil is semi-permanently flooded and the standing water is shallow, these marshes are dominated by common cattails. The marshes are often beside open water, such as a lake. Red-winged blackbirds may nest among the cattails, and beavers may build lodges nearby. Snapping turtles, water snakes and various frogs, including bullfrogs and green frogs, live here. Skunk cabbage may bloom in the early spring in the shallows or on adjacent land.

Successional old field

When farm land or cleared fields are allowed to grow free, grasses and wildflowers are typically the first plants to take root, followed by smaller shrubs and trees and then finally larger trees that may shade out and eliminate the original undergrowth. Wildflowers found here include New England asters, dandelions and Queen Anne's lace. Animals seen here include eastern cottontails, meadow voles, field sparrows, red foxes, woodchucks and white-tailed deer. Successional old fields occur statewide.

Northern hardwood forest

These forests occur mainly in the northern third of Pennsylvania. Trees found here include sugar maples, yellow birches, American beeches and black cherries. Mammals found here include white-tailed deer, black bears, porcupines, eastern chipmunks and red squirrels. Birds found here include black-capped chickadees, blue jays and cedar waxwings. Because of the shading by trees, many wildflowers, such as wood anemones, bloom in the early spring before leaves appear.

Floodplain forest

These are forests found along almost any river or large stream in Pennsylvania that floods regularly. The tree species found in them typically include silver and black maple, cottonwood, butternut, sycamore, black willow and ash. Plants seen here include jewelweed, Virginia creeper and poison ivy. Birds include red-bellied woodpeckers, cerulean warblers and tufted titmice. Great blue herons may nest in these forests. Amphibians include Jefferson salamanders and leopard frogs.

Tidal river

This is a river that flows into the ocean so that there is some mixing of fresh and salt water and some rise and fall of the water level due to ocean tides. Fish found here include striped bass, American shad, rainbow trout, catfish and spottail shiners. An example of a tidal river is the Delaware River in southeastern Pennsylvania, where tides occur as far north as Trenton, N.J.

Confined river

This is the portion of a stream or river where pools, riffles and flat-water runs occur and where there might be waterfalls and springs. The water is usually fast-flowing and clear. Fish found here include creek chub, common shiners, darters and minnows. Rainbow trout, brown trout and smallmouth bass may have been introduced. Plants found here include water-weed and pondweed.

Urban vacant lot

Vacant lots in towns and cities can become home to a range of wildlife. Trees found here include tree-of-heaven, white mulberry and Norway maples. Wildflowers found here include Queen Anne's lace, dock and goldenrod. Birds found here include starlings, house sparrows and pigeons. Mice and chipmunks may also be seen here, and cats may prowl at night, looking for a meal.

Water lily aquatic wetland

Found throughout the state, these ecosystems are often associated with ponds, lakes and slow-moving sections of rivers. Plant species seen here include fragrant water lily and spatterdock. Green frogs, bullfrogs and dragonflies are common, and birds seen here may include great blue herons and mallards.

Appalachian oak forest

This is a forest mainly of hardwood trees that is usually found on ridgetops and slopes. Trees found here include red and white oaks, red maples and hickories. Oaks produce acorns, so animals seen here include those that feed on acorns, such as gray squirrels, deer mice and wild turkeys. Birds seen here include hawks that feed on small mammals, as well as blue jays and tufted titmice. Timber rattlesnakes may also be found here.

A brown trout, bottom, and a northern pike

Bluegill – 4 to 12 in., typical length for adults – fresh water

White perch – 8 to 10 in. – fresh and salt water

Smallmouth bass – 8 to 15 in. – fresh water

Rainbow trout – 10 to 25 in. – fresh and salt water

Striped bass – 18 to 45 in. – fresh and salt water

This planet was made for fish. Nearly 71 percent of the Earth's surface is covered by the oceans. Add in the countless lakes, ponds, rivers and streams on land and it's clear that uncounted billions of fish call all this water home.

In Pennsylvania, there are more than 160 species of fish that can be found in the waters of the state's lakes, ponds and rivers. One can find bigmouth buffalo and white bass in Lake Erie, yellow perch and bullheads in almost any large lake or pond, as well as striped bass and muskies in the Delaware and Susquehanna rivers.

Some fish, such as the yellow perch and bluegill, are found only in freshwater habitats. However, American shad spend most of their time out at sea in salt water, then they come up rivers, including the Delaware and Susquehana rivers, to spawn (reproduce) in the spring in freshwater environments.

Fish evolved well before land animals, first appearing nearly 500 million years ago. Most scientists believe land animals can claim fish as their ancestors. They believe amphibians, such as salamanders and frogs, evolved directly from fish; reptiles, such as snakes and turtles, evolved from amphibians; and birds and mammals, including human beings, evolved separately from reptiles.

Fish breathe underwater by drawing oxygen from water as it passes over their gills. Sharks are fish, but whales (including dolphins and porpoises, which are kinds of whales) are mammals and they breathe air into their lungs like human beings, holding their breath when they submerge. Some whales can stay submerged for more than an hour on one breath.

Most fish are among those animals whose reproductive strategy is to lay lots of eggs and then abandon them. Enough eggs usually develop that their species will survive. Fish usually produce thousands or even millions of eggs each season.

Pennsylvania fish and game workers have been stocking trout and other fish in the state's inland waters since the late 1800's. Back then, young trout were raised in state hatcheries, loaded into milk cans filled with water, and transported to the release site by horse and buggy or by train.

Pennsylvania has nearly 4,000 inland ponds and lakes, many of which are very isolated and have never been stocked. Yet, fish somehow manage to find their way into these waters.

How does it happen? Of course, many ponds are connected to others by streams, but some are cut off entirely from other ponds and lakes with fish. Wildlife biologists believe birds may have carried fish eggs in their feathers or on their feet or even in their droppings from one pond to another. Also, violent storms, floods and even tornadoes may have moved fish or their eggs from one pond to another. Such things only have to happen once in a thousand years for a fish population to get started in a pond or lake.

Trophy fish

Here are the record weights, in pounds and ounces, as of 2015 for some species of fish caught in Pennsylvania waters.

White perch – 1 lb. 12 oz.
Bluegill – 2 lb. 9 oz.
Yellow perch – 2 lb. 11 oz.
Crappie – 4 lb. 2.88 oz.
Brook trout – 7 lb.
Smallmouth bass – 8 lb. 8 oz.
Chain pickerel – 8 lb. 14.8 oz.
American shad – 9 lb. 9 oz.
Largemouth bass – 11 lb. 3 oz.
Rainbow trout – 15 lb. 6.25 oz.
Walleye – 17 lb. 9 oz.
Brown trout – 19 lb. 10 oz.
Chinook salmon – 28 lb. 15 oz.
Lake trout – 29 lb. 4 oz.
Northern Pike – 35 lb.
Channel catfish – 35 lb. 2.5 oz.
Common carp – 52 lb.
Striped bass – 53 lb. 13 oz.
Muskellunge – 54 lb. 3 oz.

In weather not fit for man or beast, the beasts and other forms of wildlife manage to do a pretty good job of surviving in winter.

Through the ages, plants and animals have gained a variety of tricks for getting through even the most severe winter weather.

Some animals hibernate, sleeping through the worst of it. Others know how and where to avoid biting winds and freezing temperatures and where to find food in a pinch.

Birds can seem the most vulnerable of animals in winter, but they often ride out storms perched high up inside the layered branches of dense evergreen trees, sheltered from the snow and wind, sleeping or eating insects they find in the bark and under limbs. Their feathers offer natural protection from the cold. They can also fluff up their feathers with air for even greater insulation.

In winter, some birds will form great flocks, sometimes thousands of birds, and spend the night together in trees or other suitable locations, partially for warmth and partially for protection from predators and cold winter winds. Streams of crows can often be seen around dusk in the evening moving to their nightly roost.

Perhaps the largest crow roost in the state forms in the Hill District of Pittsburgh. As many as 40,000 birds have been counted there on some nights.

Woodchucks and bats hibernate through the winter.

Raccoon

Red fox

Winter survival strategies

White pine
The needles of evergreens, such as pines, are not very sensitive to cold because of molecules within their cells that act like antifreeze in a car. Also, the cells contain low amounts of water.

Day lily
Some plants, like corn, die completely at the first frost, leaving their seeds behind to produce new plants in the spring. In others, like day lilies, the parts above ground die when winter arrives, but underground roots and stems live to grow again in the spring.

Sugar maple
Leaf-bearing trees, such as maples, shed their frost-sensitive leaves in the fall and move a lot of their water and nutrients – their sap – into their roots, away from cold temperatures.

Honey bee
Honey bees form large clusters inside their hives and create heat by shivering or beating their wings. To prevent the bees on the outside of the cluster from freezing to death, those on the inside change places with them from time to time.

Beavers spent the winter awake in their lodges. Rabbits take refuge in tunnels under snow-covered shrubs or in abandoned woodchuck holes. For these and other mammals, their fur is their winter coat.

For coyotes, red foxes, bobcats and a variety of birds of prey that choose not to migrate, such as red-tailed hawks, winter in this region is business as usual as they prowl and patrol the countryside for a meal. Snow can actually benefit them since it can make prey stand out.

For both plants and animals, the greatest danger of severe cold is that it can freeze water. When water does freeze, it expands and forms sharp-edged ice crystals. The cells of all living things are filled with water, and if that water turns to ice, it can puncture the cell walls, causing damage or even death.

Perhaps the most fantastic trick for surviving freezing temperatures is the one used by some frogs. Many frogs spend the winter on land, buried beneath leaves on the forest floor. But it may become so cold that even these places have freezing temperatures. To cope with such conditions, treefrogs, wood frogs and even spring peepers have found a way to turn to ice in winter, their bodies nearly frozen solid, and still survive. With just a few warm spring days, they will thaw out and begin to do the things frogs normally do in the spring.

How do they do it? They create chemical "seeds" within their body cavities but outside the cells of their organs and other body tissues. When the temperature drops below freezing, ice crystals will form around the seeds so that the water inside the cells does not freeze. Up to two-thirds of a frog's body water can freeze, yet it can remain alive because its cells are unharmed.

Cottontail rabbit

Mallards

Hibernation

While many animals sleep through much of the winter, not all hibernate. True hibernation involves a dramatic decline in both the heart rate and the body temperature. For instance, the heart rate of a woodchuck falls from 100 beats per minute when active to 15 during hibernation, and its body temperature falls from 96 degrees to 47 degrees Fahrenheit. Snakes, turtles and frogs hibernate, but among mammals in Pennsylvania, the true hibernators are the woodchuck, jumping mouse, chipmunk (although it will emerge from sleep to eat periodically) and several species of bats.

Black bears are not true hibernators. Their body temperature does not fall significantly. Skunks and raccoons will sleep through the winter but they do not hibernate. Birds that remain in the area during winter do not hibernate, but they do sleep. However, it is a different kind of sleep than the sleep of mammals. It occurs in shorter cycles. Also, half s bird's brain remains awake during sleep and it can control how deeply the other half sleeps.

It's not called the urban jungle for nothing. A city can be teeming with animals, from birds to bats to butterflies, and people who live there may be unaware of most.

That's because many animals that do choose to live where there are great concentrations of people eventually learn ways to live their lives without attracting notice.

There are some animals, though, that have the personality – and the agility – to live among people but to stay out of their way. Gray squirrels and pigeons can seem to be acrobats of city sidewalks as they avoid being trampled.

Still other animals thrive in cities because they are terrific survivors. Norway rats and German cockroaches, two species people especially dislike, have managed to survive despite the most intense efforts imaginable to eliminate them.

But even in urban areas, there can be rural landscapes in the form of city parks. They are often havens for animals. For birds especially, a park's greenery can be an oasis in the middle of a desert of concrete and steel. In fact, concentrations of some birds can be higher in a good-sized park than they are in the outlying rural areas.

That's especially true in spring and fall when birds migrate. They have to stop and rest periodically, and if they are flying over a densely populated area with lots of buildings and little greenery, a park will stand out for them like the bull's-eye of a target. That's why Fairmount Park, the municipal park system of Philadelphia, with nearly 9,200 acres, is such a prime bird-watching destination for people. Some 231 species of birds have been seen in the park system, with many of them spotted only during the migration periods.

Suburban towns can also harbor vast populations of wild animals. Some animals, such as deer, skunks, raccoons and opossums, are attracted by the easily obtained

House sparrow nest

food provided by bird feeders, vegetable gardens, compost piles and trash cans. The breeding season will also send animals into areas where they are not usually found – primarily males, which may wander widely in search of females. And because hunting is not allowed within many cities and towns, wildlife may thrive here and even learn to feel safe.

Black bears have become an increasing problem, as their population in Pennsylvania has more than tripled in the last 30 years. Hungry when they come out of hibernation, and with little food yet available in the wild, they may invade residential areas in search of bird feeders and trash cans with little feeling that they are in danger. In 2014, after being spotted wandering around northeast Philadelphia, a bear was captured just a few miles from the city limits. In 2012, a young black bear entered a suburban mall outside Pittsburgh through the automatic doors, exploring the mall for nearly an hour before being captured.

For some animals, the feeding opportunities in cities can be greater than in the deep forest. Peregrine falcons were originally a mountain species that vanished from the state due to use of the pesticide DDT, which was eventually banned. They would build their nests on cliff ledges and use that vantage point to spot prey, usually other birds, which they would attack in flight.

Falcons have since returned, but now they find the man-made cliffs of cities serve them just as well. They will nest on building ledges or on the girders of bridges and then hunt the canyons of the city for pigeons and other birds. In Pennsylvania, they have established nests on the Ben Franklin Bridge in Philadelphia, the Eighth Street Bridge in Allentown, the Gulf Tower in Pittsburgh, the Fulton Bank in Harrisburg and the Callowhill Building in Reading.

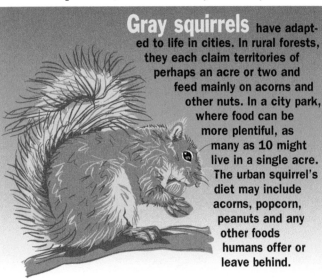

Gray squirrels have adapted to life in cities. In rural forests, they each claim territories of perhaps an acre or two and feed mainly on acorns and other nuts. In a city park, where food can be more plentiful, as many as 10 might live in a single acre. The urban squirrel's diet may include acorns, popcorn, peanuts and any other foods humans offer or leave behind.

Peregrine falcon

Delaware River as it passes through the Delaware Water Gap National Recreation Area

Impressionist painters would not have gotten far without wildflowers.

Wildflowers were art before there was art – dashes of color on canvases of windblown meadows, craggy mountainsides and darkened forest floors.

Many seem to have been named by poets – Jack-in-the-pulpit, Queen Anne's lace, butter and eggs, yellow lady's slipper.

But as unplanned as a scattering of wildflowers may seem, where and when they bloom and the colors and shapes they take follow a general logic. Much of that logic has to do with helping each species of wildflower find its own place in the world.

If all wildflowers bloomed at the same time, grew in the same soil and sunlight conditions and attracted the same pollinators, the fierce competition would mean that only a few species would survive. So wildflowers evolved to be specialists. Some grow only in wet areas, some grow only in dry areas. Some thrive in valleys and others on mountainsides. Some can grow where there is a lot of sunlight, as in a meadow, while others can grow where there is little of it, as on a forest floor.

To further reduce competition, wildflowers also evolved to bloom at different times of year. Common violets bloom from April to June, but most asters don't bloom until September or October.

The beauty of wildflowers is not something created to please the human eye. It actually evolved to appeal to pollinators – butterflies,

Blue flag iris

Wild lupine

bees, ants, moths and even small birds, such as hummingbirds. The splashy and bold colors of some wildflowers are like neon signs outside a row of stores that aim to catch the eyes of passing customers. The fragrances the flowers give off and the sweet nectar produced inside the flowers are also intended as attractions to pollinators.

A wildflower must get the pollen that the male part of the flower creates to the female part of another wildflower of the same species so that seeds can be created. Pollinators are the pollen carriers.

If a butterfly visits a flower to feed on the nectar deep inside the flower, tiny grains of pollen may stick to it and be carried to the next flower it visits.

The varied shapes and colors of wildflowers also evolved to let a pollinator know exactly what kind of flower it is visiting. Most pollinators can recognize certain shapes and patterns of color, so if they like the nectar they find at a flower, they will know to visit that same type of flower again. In this way, pollen from a buttercup has a way of getting to another buttercup, rather than ending up on a wild geranium.

The bull's-eye shape of many flowers evolved to guide the pollinator to the nectar. "Here it is – right at the center," this shape seems to say. Other wildflowers have a shape like the end of a trumpet for the same reason. "The nectar is right inside here – you can't miss it," this shape seems to say.

In Pennsylvania, there may be more than 1,500 species of wildflowers that can be seen through the seasons.

Wildflower or weed?

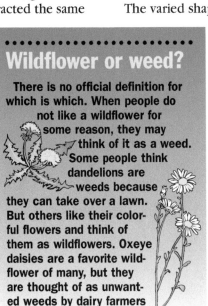

There is no official definition for which is which. When people do not like a wildflower for some reason, they may think of it as a weed. Some people think dandelions are weeds because they can take over a lawn. But others like their colorful flowers and think of them as wildflowers. Oxeye daisies are a favorite wildflower of many, but they are thought of as unwanted weeds by dairy farmers because cows may refuse to graze on them.

Turk's-cap lily

Red clover

Fragrant water lily

CHECKLIST
Common wildflowers

Oxeye daisy
June – Aug.
1 – 3 in. wide
☐

Daisy fleabane
May – July
1 in. wide
☐

Queen Anne's lace
June – Oct.
2 – 6 in. wide
(cluster width)
☐

Fragrant water lily
June – Sept.
3 – 6 in. wide
☐

Tall goldenrod
Aug. – Oct.
.2 in. long
☐

Common dandelion
April – Oct.
1.5 in. wide
☐

Black-eyed Susan
June – Oct.
2 – 3 in. wide
☐

Common buttercup
May – June
1 in. wide
☐

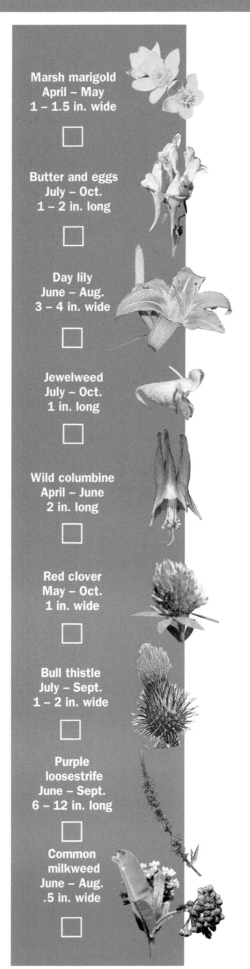

Marsh marigold
April – May
1 – 1.5 in. wide
☐

Butter and eggs
July – Oct.
1 – 2 in. long
☐

Day lily
June – Aug.
3 – 4 in. wide
☐

Jewelweed
July – Oct.
1 in. long
☐

Wild columbine
April – June
2 in. long
☐

Red clover
May – Oct.
1 in. wide
☐

Bull thistle
July – Sept.
1 – 2 in. wide
☐

Purple loosestrife
June – Sept.
6 – 12 in. long
☐

Common milkweed
June – Aug.
.5 in. wide
☐

Common blue violet
April – June
.5 – 1 in. wide
☐

Pickerelweed
July – Oct.
4 – 6 in.
☐

Spotted Joe-pye weed
Aug. – Sept.
4 – 5 in. wide
☐

Wild lupine
May – July
.7 in. long
☐

Wild geranium
May – June
1 – 2 in. wide
☐

Smooth aster
Sept. – Oct.
1 in. wide
☐

Skunk cabbage
March – May
3 – 6 in. long
☐

Cattail
June – Sept.
6 in. long
☐

Jack-in-the-pulpit
May – June
3 – 4 in. long
☐

Showy lady's slipper

Wild columbine

Violets

Spotted touch-me-not

Butter and eggs

Mushrooms – A fungus among us

There is a saying: "There are old mushroom hunters and bold mushroom hunters, but there are no old, bold mushroom hunters."

After a warm rain, lawns can turn into bountiful mushroom factories, as puffballs, shaggy manes and boletes sprout seemingly overnight. One may be tempted to look on these fungal colonies (a mushroom is indeed only an elaborate fungus) as a free source of something you would have to otherwise pay for in a supermarket. But experts warn against it, given that so many are poisonous.

There may be several thousand species of mushrooms to be found in Pennsylvania, including some that are deadly. There may be 10,000 species of mushrooms growing throughout North America, of which about 250 are considered edible, and about the same number are known to be poisonous, sickening or even capable of killing those who eat them. The rest are, to varying degrees, still mysteries.

Among the poisonous mushrooms, there are about 10 North American varieties that are considered deadly. And among those, the most feared is Amanita phalloides, the "death cap," which, although rare, has been found growing in suburban yards in the Northeast and elsewhere.

Its name is not hype. Eating just an ounce can be fatal.

Compounding the problem is that there are some poisonous mushrooms that can look very much like edible varieties. For instance, the destroying angel, which is deadly poisonous, can look almost identical to a type of common meadow mushroom.

The rule of thumb is that unless you know it, don't eat it.

Death cap mushroom, 3 in. tall

Orange oak bolete

Reg-veil amanita

Fly agaric

Ling chih

Russula Emetica

Ozone Falls in Ricketts Glen State Park

Although Pennsylvania is the 33rd largest state, only Alaska and California have more state park land. There are 120 state parks in Pennsylvania, totaling nearly 300,000 acres, almost one percent of all the state's land.

In line with the commonwealth's intention of having a state park within 25 miles of every resident, there are state parks in 61 of the state's 67 counties. They range in size from three acres (Sand Bridge State Park in Union County) to 21,122 acres (Pymatuning State Park in Crawford County).

Fittingly, the first Pennsylvania state park was at Valley Forge, the site of the 1777-78 winter encampment of George Washington's Continental Army during the American Revolution. The park was donated to the National Park Service as part of the nation's bicentennial in 1976.

In addition to state parks, Pennsylvania also has 20 forest districts that cover an estimated 2.2 million acres of land, which means those forest lands are permanently protected.

Some notable state parks

Ricketts Glen S.P. in Luzerne, Sullivan
and Columbia counties is 13,050 acres. It boasts 22 waterfalls and 26 miles of hiking trails as well as a 600-foot beach and the 245-acre Lake Jean for fishing. It also contains the Glens Natural Area, a National Natural Landmark known for its old-growth forests and its numerous waterfalls along Kitchen Creek.

Ohiopyle S.P. in Fayette County is about
20,500 acres. While it has 79 miles of hiking trails, 27 miles of biking trails and spectacular scenery, especially along the Youghiogheny River, the centerpiece of the park is the 14 mile-long Youghiogheny River Gorge. It offers some of the best whitewater rafting, kayaking and canoeing in the East.

Presque Isle S.P. in Erie County is 3,200
acres. Located on a sandy, curving penninsula that sweeps out into Lake Erie, the park has 11 miles of hiking trails and more than a dozen beaches (with lifeguards) for swimming. In winter, Presque Isle Bay is popular for ice fishing and ice skating, and the park's trails are open for cross-country skiing.

Worlds End S.P. in Sullivan County is 780
acres. It occupies the narrow steep valley of Loyalsock Creek and is surrounded by the 115,000-acre Loyalsock State Forest. The park is known for its rugged natural beauty and the spectacular views from some of its hiking trails (more than 20 miles of them) and scenic overlooks, which can be reached by car.

Parks and gardens

The greenery of a park is as necessary to the vitality of a city as fresh air and sunshine are to the health of its residents.

Indeed, the greenery and other color provided by parks as well as street trees, window boxes, median planters and private landscaping all enhance the sense of nature amid the city.

Philadelphia has one of the oldest and largest systems of urban parks in the United States. The Fairmount Park system contains 63 parks covering 9,200 acres. The largest park in the system, at 4,100 acres, is its namesake, Fairmount Park along the Schuylkill River. It is also home to the world famous Philadelphia Zoo, the first zoo in the United States.

Bringing farming into the city has been the accomplishment of community gardens, an idea that first took root in the United States in Detroit in the 1890's as a way to give the unemployed something to do.

A community garden

The idea caught on and other cities soon began similar programs. However, during World War I, such gardens became an important source of fruits and vegetables in a time when there were severe food shortages across Europe. During the Great Depression, when money became scarce, growing your own food became a necessity for many. And During World War II, "victory gardens" again became a way to make up for widespread food shortages. In 1944 alone, such gardens provided 42% of the nation's vegetables.

In the 1970's, there was a revival of interest in community gardens rising out of two popular movements – environmentalism and urban activism. Today, it's estimated there are nearly 18,000 community gardens in North America.

Vacant lots as accidental gardens

A patch of earth in a city landscape is a window of opportunity for nature.

Examine any recently cleared vacant lot in a city, and then look at it a year or two later. Completely ignored, subjected to litter and pollution, given no ration of water but what the skies provide, this forgotten piece of property is likely to become a teeming garden of wildflowers and other plants in that time, an isle of greenery amid a sea of concrete and asphalt.

Typically, each spring and summer dozens of plants will manage to grow and flower there, revealing a spectrum of colors through the seasons – among them butter and eggs, Queen Anne's lace, phlox, spurge, dock, yarrow, clover, thistle, mullein, bittersweet, peppergrass, milkweed, ragweed and primrose.

In these urban lots, plants tend to stake out their own territories – phlox and peppergrass over here, thistles and spurge over there, almost as if they were deliberately planted in sections. That's often nature's pattern, a characteristic of a site that is relatively young. When a lot is first cleared, one plant will get established and the seeds will drop around it, so that it will spread out from there, colonizing the area immediately around it. A little farther away, another plant will be doing the same thing.

A vacant lot

In the first year, you will typically find plants popping up like mullein and Queen Anne's lace. They don't need a season of cold before they germinate. That's one of the reasons they are so prolific and they become established so quickly.

How do plants find their way into these isolated urban sites in the first place? It's because their seeds are expert travelers. Some arrive in bird droppings or when birds drop them in flight. Butterflies and other pollinators also bring some in, and many are blown in by the wind. In some cases, people carry them in. They may rub against a plant, such as milkweed or thistle, and the seeds might stick to their clothes and later drop to the ground.

Eventually, tree saplings will take root, species such as trees-of-heaven or mulberry trees. Soon, the growing trees will shade out many sun-loving species of ground plants and a different group of shade-tolerant plants will find their way into the mix. And, as often happens in an urban area, city workers may eventually cut these "weed trees" before they grow so tall that their roots buckle the nearby sidewalks or streets. They may even clear the plant growth. Then the lot is back to being a blank canvas on which nature is always willing to paint.

Penn's woods. Roughly translated, that's what Pennsylvania means. And at the time William Penn founded the colony of Pennsylvania that later became the state, Penn's woods was an accurate name. In the late 1600's, nearly 95 percent of the state was forested.

However, the population steadily increased, land was cleared for farming, industry and homes, and trees were harvested for building and other purposes. By 1900, only 32 percent of the state was still forested. And with forestland vanishing nationwide, a conservation effort picked up steam to save forestland.

At the same time, people were moving from rural to urban areas, farms were being abandoned and logging in Pennsylvania was declining. States, including Pennsylvania, began buying forested land to protect it permanently, and trees were planted on abandoned farms and elsewhere to rebuild the forests.

Today, nearly 60 percent of the state, some 17 million acres, is forested. And the state owns and protects 2.2 million acres of that forestland, roughly 13 percent.

Surely, there are billions of trees in the state – too many trees to possibly count. Not true. Just as it takes a census of people periodically, the federal government also takes an inventory of trees. In 2013, during the latest such inventory, the Forest Service estimated there were nearly 8.3 billion trees at least an inch in diameter at breast height in Pennsylvania. Red maples make up nearly a quarter of all trees.

Just because so much of Pennsylvania's trees are protected does not mean they are completely safe, though. Over the years, tree diseases and insects have dramatically altered the state's landscape. At one time, a quarter of the trees in the state were chestnuts, highly valued for their wood and nuts. But chestnut blight, a type of fungus, swept through Pennsylvania forests in the early 1900s, virtually eliminating them.

Gypsy moths, insects that eat tree leaves, particularly those of oaks, have also done considerale damage. In one year alone, 1990, these moths stripped the leaves from nearly a quarter of the trees in the state, killing many.

The greatest current threat to forestland may be the exploding population of white-tailed deer, some 1.6 mil-

lon currently. Deer feed on the seedlings of many trees, preferring those of oaks, birches, ashes and maples. So in some forests, there are few young trees growing on the forest floor, readying to take the place of older trees as they age and fall, clouding the future of these forests.

Dominant forest types of Pennsylvania
Main tree species by zone

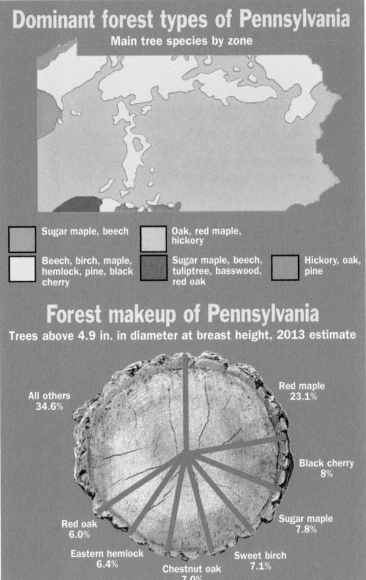

- Sugar maple, beech
- Beech, birch, maple, hemlock, pine, black cherry
- Oak, red maple, hickory
- Sugar maple, beech, tuliptree, basswood, red oak
- Hickory, oak, pine

Forest makeup of Pennsylvania
Trees above 4.9 in. in diameter at breast height, 2013 estimate

All others 34.6%
Red maple 23.1%
Black cherry 8%
Sugar maple 7.8%
Sweet birch 7.1%
Chestnut oak 7.0%
Eastern hemlock 6.4%
Red oak 6.0%

Is fall foliage just an accident?

Perhaps nature's most brilliant artistic touch, the scarlets and golds of autumn leaves may also be one of nature's most beautiful accidents.

The startling colors of fall are certainly admired by human beings. But the colors seem to defy one of nature's primary laws, since they are of no use to the trees themselves.

Most plants and animals have the features they have because those features help them survive in the world. But what use is a colorful leaf to a tree in autumn since the leaf is just about to die and fall

to the ground?

Some scientists believe that autumn colors may have begun as an accident, the unexpected result of the chemical changes a tree goes through as it prepares to lose its leaves each fall.

But even if it was an accident, since this coloring causes no harm to the trees, there is no reason for trees to lose this process. If the coloring did cause harm, those species of trees whose leaves changed color might have become extinct by now, and only the trees whose leaves didn't change color might be alive today.

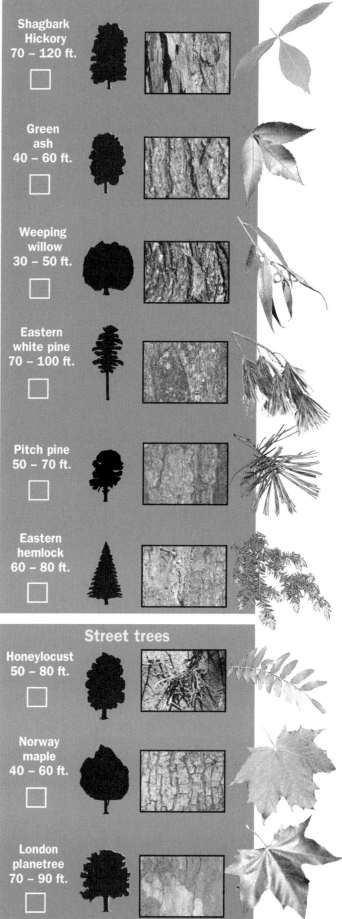

CHECKLIST
Common trees

Sugar maple
typical height
60 – 80 ft.
☐

Red maple
50 – 80 ft.
☐

Northern red oak
60 – 80 ft.
☐

Chestnut oak
60 – 80 ft.
☐

Sweet birch
60 – 80 ft.
☐

Paper birch
50 – 70 ft.
☐

Black cherry
50 – 70 ft.
☐

American beech
60 – 80 ft.
☐

Shagbark Hickory
70 – 120 ft.
☐

Green ash
40 – 60 ft.
☐

Weeping willow
30 – 50 ft.
☐

Eastern white pine
70 – 100 ft.
☐

Pitch pine
50 – 70 ft.
☐

Eastern hemlock
60 – 80 ft.
☐

Street trees

Honeylocust
50 – 80 ft.
☐

Norway maple
40 – 60 ft.
☐

London planetree
70 – 90 ft.
☐

Apple

Red maple

Sugar maple

Eastern white pine

Norway Spruce

Seed science

The apple doesn't fall far from the tree, the saying goes. However, for a species of tree to survive and spread, there has to be a way for the seeds to travel to new locations.

The seeds of an apple tree are contained in the apple and when animals eat the lucious fruits, they inadvertently provide a way for the seeds to travel to a new place – usually in their digestive tract. When the animal finally passes solid wastes, the seeds encased in a hard shell end up on the ground, ready to take root, amid ready-made fertilizer.

Some trees, like red and sugar maples, produce winged seeds that travel on the wind to distant places like miniature helicopters. Other trees, like American beeches, produce tasty nuts (tasty to humans also) that are seeds as well. The acorns of oak trees are also seeds.

Coniferous trees, like pines, hemlocks and spruces, produce edible cones with the seeds inside. When a squirrel or other cone-eating animal carries the cone a distance to eat it, the seeds have a chance to travel too.

Cones, seeds and nuts are shown their actual size.

Pitch pine

Eastern hemlock

American beech

Northern red oak

Yellow birch

An acorn from a northern red oak pushes out a taproot

An acorn that drops to the ground in autumn will grow a main root, called a taproot, that pushes down into the soil once temperatures warm in the spring. Within a week the root may be several inches long. Within another week, a stem may begin to grow upward and leaves will soon appear on the stem.

Because so many acorns are eaten or do not fall in a place where they can germinate, it may take 500 acorns to produce one oak seedling that reaches a year old.

The tiny acorns that litter the ground on hiking trails in the fall can be the key to survival for many forms of wildlife in a forest ecosystem. After all, mighty oaks from little acorns grow. And on little acorns a range of animals feed, including black bears, deer, turkeys, squirrels, chipmunks, mice, blue jays and vaious insects. And many of the animals that don't feed on acorns feed on animals that do. The nut-eating animals are often prey for larger animals, such as foxes, coyotes, bobcats, hawks and owls.

However, if all the acorns that fell in autumn got eaten, there would be no mighty oaks, and if there were no mighty oaks, the populations of all those animals that depend on acorns, either directly or indirectly, would be in trouble. So oaks have evolved to have a fascinating strategy that ensures that both oaks and animals survive.

An acorn still on the branch

Oaks have years when they produce a huge crop of acorns, called mast years. For oaks, mast years occur every three to five years on average. But in the years between mast years, oaks may produce few if any acorns. That way, the populations of the acorn-eating animals have less food in the in-between years and can never grow so large that all the acorns get eaten in the mast years. That guarantees new oaks will be able to take root while allowing the animals that depend on acorns to survive.

But here is the fascinating part. If one oak had a mast year and the oak beside it did not, this process would not work. Acorn eaters would still have enough food that their populations would grow without limit, and in mast years, all the acorns might get eaten. But that doesn't happen.

In one of the great mysteries of nature, oaks, and other nut-bearing trees, have evolved so that somehow mast years occur at the same time for trees of that species over a vast area, sometimes thousands of square miles. That means the population of the oaks and the animals that eat acorns and other nuts remain in balance, and all have a chance to survive.

Oaks can't communicate with each other, as far as anyone knows. So how do they all agree to grow so many acorns at the same time? One leading theory is that variations in seasonal temperature, especially in the spring, trigger trees over a large area to produce large or small crops of nuts. One study of oaks in California found that in warm, dry Aprils, acorn production tended to be high in the fall.

They are the silent observers of history. Pennsylvania's oldest living trees were alive when Native Americans dominated the region, when the Revolutionary War was fought, and when electric lights first appeared.

Indeed, there may be 10,000 or more acres of old-growth forests in the state. The oldest tree in the state and the second oldest in the Northeast is an eastern hemlock in Tionesta in Forest County. It is believed to be 555 years old.

There are also tall trees in the state. One, a white pine tree, called the Longfellow pine, in Cook Forest State Park in Clarion County is the tallest tree in the Northeast at nearly 185 feet.

How do scientists determine the age of a tree? By counting its rings. A tree grows new wood around the outside of its trunk each year. In the late spring and early summer, when there is usually more moisture in the ground, the wood is lighter in appearance. In late summer and early fall, when conditions are usually drier, the wood being added tends to be darker. This alternating light and dark wood creates rings that can be seen when the tree falls in a storm or is cut.

To count the rings of a living tree, a scientist drills out a pencil-thin "core," a rod of wood extracted from the trunk on which the rings can be seen.

There is no firm definition for an old-growth tree, but for many species, living 150 to 200 years can put them in that category. Among the long-lived species in Pennsylvania are eastern hemlocks, eastern white pines, white cedars, balsam firs, red spruces, sugar maples and American beeches.

In Pennsylvania, much of the land was cleared for farming in Colonial times, even the lower slopes of mountains, and later, many forests were heavily logged. So most of the old-growth stands of trees that exist today are found in the higher elevations of mountains or in places that people can't easily reach, such as on steep slopes, in ravines or on remote plateaus.

One of the largest parcels of old-growth trees in the state, some 4,000 acres, is in the Allegheny National Forest in northwestern Pennsylvania. On a pleatau that escaped being heavily logged, the site is filled with American beeches, eastern hemlocks and sugar maples.

While 500 years' longevity is impressive for Pennsylvania, it wouldn't raise an eyebrow in some other regions. There is a bald cypress in North Carolina that is estimated to be 1,622 years old, the champion for old growth in the eastern United States.

The champions for the nation, and perhaps the world, are bristlecone pines that grow on the barren windy slopes of the southwestern United States. Some of those may be nearly 5,000 years old.

Eastern white pine

Old-growth trees of Pennsylvania
Some major stands

McKean County
In the Allegheny National Forest, in the Tionesta Scenic and Research Natural Area, nearly 4,100 acres of eastern hemlocks, American beeches and sugar maples

Forest and Clarion counties
In Cook Forest State Park, perhaps 2,300 acres of red and white oaks, red maples and black cherries

Centre County
In the Bear Meadows Natural Area, about 320 acres of black spruces, balsam firs, eastern hemlocks and yellow birches

Luzerne County
In Ricketts Glen State Park, nearly 2,000 acres of eastern hemlocks, white pines and oaks

Monroe County
Near Pocono Lake, perhaps 5,000 acres of pine barrens with old-growth pitch pines and scrub oaks

Snyder County
In the Snyder Middleswarth Natural Area, some 250 acres of eastern hemlocks, white pines and pitch pines

Tall trees of Pennsylvania
By species

Black cherry – In Cook Forest State Park, 137.3 feet, tallest in the Northeast

White pine (called the Longfellow pine) – In Cook Forest State Park, 184.7 feet, the tallest tree of any species in the Northeast

White oak – In Cook Forest State Park, 127.3 feet, tallest in the Northeast

Red oak – In Fairmount Park in Philadelphia, 135.2 feet, tallest in the Northeast

Eastern hemlock – In the Snyder Middleswarth Natural Area in Snyder County, 145.3 feet, tallest in the Northeast

American chestnut – In Cook Forest State Park, 84.2 feet, tallest in the East.

Red maple – In Wintergreen Gorge in Erie, 136.6 feet, tallest in the Northeast.

Snowy egrets in their nest

Imagine mountains as high as Mount Everest. Imagine erupting volcanoes. Imagine mile-high glaciers or frozen snowswept plains.

You are imagining landscapes that could have been seen in the Northeast at different times in its history.

Hard to believe? Not for geologists. They know that the appearance of every region on Earth has changed over time. They also know that now is a quiet time in the geologic cycle of Pennsylvania and the Northeast, a momentary calm amid the violent upheavals that have marked the history of the region's landscape.

The Earth's crust, or surface layer, is made up of vast rock plates that form the continents and the ocean floor. There are about a dozen large plates and many smaller ones covering the Earth's surface, arranged like pieces in a jigsaw puzzle.

The crust under the oceans may have an average thickness of just four or five miles. The crust of the continental plates is about 25 miles thick on average.

Essentially, the plates "float" on a layer of denser rock, called the mantle, deeper within the Earth. The Earth's core has a temperature of perhaps 10,800 degrees Fahrenheit, and this great heat creates circulating currents

Layers of sedimentary rock in exposed bedrock by a highway

of molten rock within the mantle, which slowly push the crustal plates along.

The plates typically move a few inches a year, sometimes coming together and sometimes moving apart. This is called continental drift.

Even at such a slow speed, though, when these massive plates push against each other, geologic fireworks happen. Mountains gradually rise into the air, volcanoes may erupt and earthquakes might rumble.

That's what's happening currently in the western United States. The Pacific and North American plates are slowly sliding by each other, grinding together along their edges. About 65 million years ago, when the plates were pushing right into each other, this process created the Rocky Mountains, which are considered "young" mountains. (Scientists believe the Earth is about 4.6 billion years old.)

Pennsylvania is also part of the North American plate, but for nearly 200 million years this plate has been separating from the plate that is next to it on the east, the one on which Africa is located. So there are no strong geologic pressures on Pennsylvania's part of the plate to create or enlarge mountains, or to produce volcanoes or sizable earthquakes, in the Northeast.

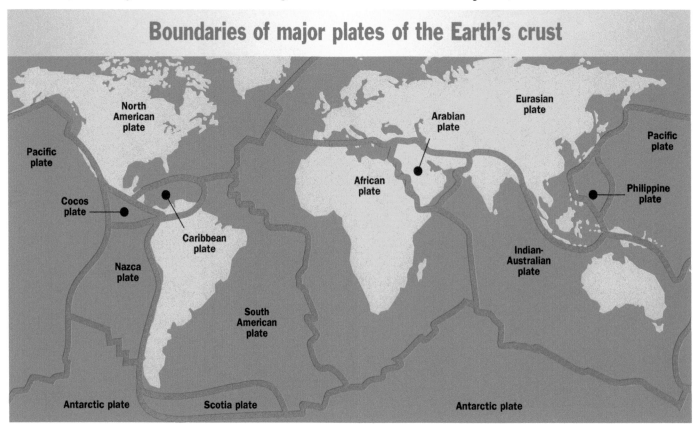

Boundaries of major plates of the Earth's crust

North American plate

Eurasian plate

Arabian plate

Pacific plate

Pacific plate

Cocos plate

African plate

Philippine plate

Caribbean plate

Nazca plate

Indian-Australian plate

South American plate

Antarctic plate   Scotia plate   Antarctic plate

Drifting continents

At least three times in the last 500 million years, the East Coast of North America collided with other continental plates. These collisions did much to shape the landscape of Pennsylvania.

Although continental plates may move only inches a year, their masses are so great and the forces pushing them along are so powerful that the process has long-lasting momentum.

Once the plates come together they may continue to push against each other for millions of years. This can create huge mountains in the region of the collision or, if the reaction of the land is violent, earthquakes or volcanoes.

500 million years ago

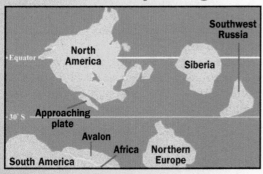

Pennsylvania is lying on a continent at the equator, but it is underwater, covered by a shallow sea. Trees, plants and animals have not yet appeared on land. Primitive life is still developing in the oceans. The first fishlike creatures appear in water. Less than a foot in length, they lack jaws but they do have backbones, perhaps the first animals that are vertebrates.

The oxygen content of the Earth's atmosphere is rising sharply, but it will be another 100 million years before air-breathing land animals evolve.

Drifting north toward early North America is a plate, covered with volcanoes, that is about to collide with what will become the East Coast.

460 million years ago

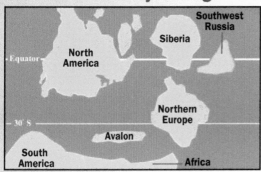

A volcanic plate collides with the North American coast, pushing up layers of rock that form the Taconic Mountains in eastern New York and western New England. The collision buckles the land that will become central Pennsylvania and covers it with volcanic ash.

Rain falling on the western slopes of the mountains creates rushing rivers that fill with eroded debris. The rivers flow out into Pennsylvania, laying down a thick wide blanket of mud, sand and gravel, rasiing portions of the state above sea level. However, by 420 million years ago, Pennsylvania is again under a shallow inland sea.

Meanwhile, Avalon, another plate, has broken free of the united African-South American continent and is drifting north.

380 million years ago

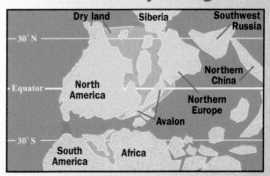

Avalon collides with North America, creating massive mountains to the east of Pennsylvania. When these peaks erode, rivers carrying debris deposit mud and sediment over the state, much of which is again covered by a shallow sea.

Primitive fish now fill the seas, and plants are starting to flourish on land.

300 million years ago

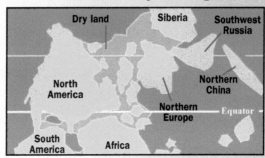

200 million years ago

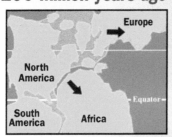

Africa collides with North America about 300 million years ago. All seven continents become joined in a single vast continent, called Pangaea by geologists. Layers of land are pushed up into Pennsylvania, adding to the height of the Blue Ridge Mountains and creating ridges and valleys in central Pennsylvania. About 200 million years ago, Africa pulls away from North America.

The first dinosaurs appear about 230 million years ago, but they are relatively small, just a few feet in height. The first mammals, not much larger than mice, appear about the same time.

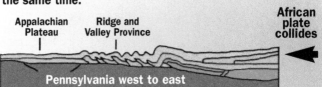

Appalachian Plateau   Ridge and Valley Province   African plate collides

Pennsylvania west to east

The rock cycle

Rocks have their own life cycle. It begins on the surface of the Earth when wind, rain and other natural forces break down existing rocks and other materials. The resulting particles become compressed and cemented together over the course of time to form new rocks.

When sand particles are turned to rock in this way, sandstone is the result. When clay, such as what is deposited on the bottom of a lake, is turned to rock, shale is the result. These two rocks are called sedimentary rocks, one of three categories of rocks along with metamorphic and igneous.

Eventually, sedimentary rocks, such as sandstone and shale, are buried by layers of other sedimentary rocks. With the heat and pressure deeper inside the Earth, they can become compressed further, making them harder still. That can turn them into metamorphic rocks, such as quartzite and slate.

Metamorphic rocks may eventually move to the surface due to earthquakes or the erosion of the land above them. But if metamorphic rocks move deeper into the Earth, the higher temperatures there can turn them to molten rock. If this molten material hardens underground, it can become igneous rock, such as granite. If it surges up to the Earth's surface through a volcano as lava and hardens, it will become another form of igneous rock, such as basalt.

Harsh weather and the freezing and thawing through the seasons eventually disintegrate surface rocks. These rock particles may then become part of new sedimentary rocks, starting the cycle once again.

The rock collection

Sandstone
Sedimentary; formed from sand

Quartzite
Metamorphic; formed from sandstone

Shale
Sedimentary; formed from mud and clay

Limestone
Sedimentary; formed from minerals such as calcite

Siltstone
Sedimentary; formed from particles of silt, clay and sand

Slate
Metamorphic; formed from shale at low temperatures

Conglomerate
Sedimentary; formed from a mix of sand and pebbles

- -
Mineral or rock?

Minerals have crystalline structures, and they are usually made up of only a few chemical elements. For instance, diamond is made of carbon, and quartz is made of silicon and oxygen. Rocks, such as schist or granite, may be made up of combinations of minerals and bits of other rocks. While the minerals in it may have crystalline structures, the overall rock usually does not.

Schist
Metamorphic; often formed from shale at medium temperatures

Dolomite
Sedimentary; formed from mud and limestone

Gneiss
Metamorphic; formed from igneous and sedimentary rocks

Quartz, a mineral

Granite
Igneous; formed when molten rock cools underground

Pennsylvania's bedrock geology

If you dig down through the loose dirt and rocks on the Earth's surface, you will eventually strike solid rock, called bedrock. In coastal areas or on steep mountain slopes, you may not have to dig deeply at all to find bedrock. In river valleys, where flooding has deposited layer upon layer of silt and clay over the centuries, you might have to dig down 100 feet or more.

Pennsylvania has distinct regions of bedrock, reflecting the different circumstances in which the rock was formed.

Much of the story of Pennsylvania's bedrock involves the uplift of huge mountains in eastern New York and western New England as the result of continental collisions, followed by the erosion of those mountains over millions of years. Rain hitting the western slopes of those mountains sent rivers filled with clay, silt, sand and pebbles out into central and western Pennsylvania, depositing that material in wide deltas. That eroded debris eventually turned into sedimentary rock.

At various times during these periods of continental collisions, Pennsylvania was covered by shallow inland seas. Along the swampy areas near the sea edges, plants would die, fall to the bottom of the swamps, where layer would build up upon layer. Eventually, these layers would become compressed and turn to coal. Deeper in the inland seas, fish and other creatures would die and their remains and shells would fall to the sea bottom, creating layers of material that eventually turned to limestone.

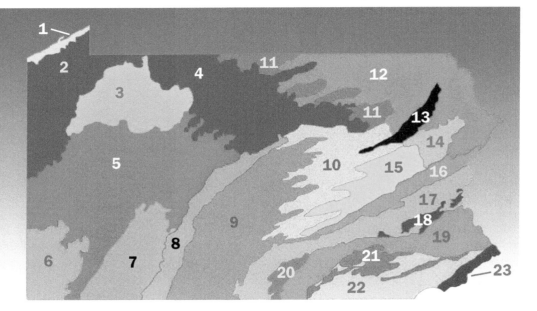

Pennsylvania's geologic zones

Underlying rock types

1 Lake Erie lowlands; shale and siltstone

2 Northwest plateau; shale, siltstone and sandstone

3 High plateau; sandstone, siltstone, shale, conglomerate and some coal

4 Plateau with deep valleys; sandstone, siltstone, shale and conglomerate

5 Pittsburgh plateau; shale, siltstone, sandstone, limestone and coal

6 Soutwest hills; shale, sandstone and limestone

7 Wide ridges separated by broad valleys; sandstone, siltstone, shale, conglomerate and some coal and limestone

8 Allegheny front; shale, siltstone and sandstone

9 Appalachian ridges and valleys; sandstone, siltstone, shale, conglomerate, limestone and dolomite

10 Susquehanna lowlands; sandstone, siltstone, shale, conglomerate, limestone and dolomite

11 Glaciated high plateau; sandstone, siltstone, shale, conglomerate and some coal

12 Glaciated low plateau; sandstone, siltstone and shale

13 Anthracite valley; sandstone, siltstone, conglomerate and coal

14 Pocono plateau; sandstone, siltstone, shale and some conglomerate

15 Anthracite uplands; sandstone, shale, conglomerate and coal

16 Parallel ridges; sandstone, siltstone, shale, some limestone and conglomerate

17 Great valley; shale, sandstone and slate in the north, limestone and dolomite in the south

18 Reading highlands; gneiss, granodiorite and quartzite

19 Gettysburg-Newark lowlands; red shale, siltsone and sandstone

20 Blue Ridge Mountains; quartzite and dolomite

21 Piedmont lowlands; limestone and dolomite

22 Piedmont uplands; schist, gneiss and quartzite

23 Atlantic Coast plain; sand and gravel with underlying schist and gneiss

Modern life depends on machines, from planes to refrigerators to computers, and the industrial revolution that brought America into modern times was largely fueled by coal, oil and natural gas that were burned to create heat and electricity.

And in this country, those fuels had their beginnings, in part, in Pennsylvania.

Anthracite coal was first discovered in the United States in 1762 at the mouth of Mill Creek on the Susquehanna River near present day Wilkes Barre by a group of settlers. In 1836, the first municipally owned natural gas distribution company was created by the city of Philadelphia. And the first commercial oil well in the United States – and the world – was drilled in 1859 by Edwin Drake in Cherrytree Township in northwest Pennsylvania.

Hundreds of millions of years ago, Pennsylvania had just the right geologic conditions to create a vast future supply of these energy-rich fuels, especially coal and natural gas.

Coal, oil and gas are forms of carbon, which is contained in living plants and animals. When plants and animals die, they fall to the ground and this dead plant matter may one day turn to these energy-rich fuels.

Some 300 to 325 million years ago, Pennsylvania was at the edge of the North American continent, which sat somewhere near the equator. Much of the state was a hot

Coal miners in Pennsylvania about 1895

swampy region where plants, such as giant ferns, reeds and mosses, grew. When these plants died, they fell into the swamps and collected at the bottom. Layer built up upon layer and these layers may have been buried eventually by sand and mud, which turned to rock over millions

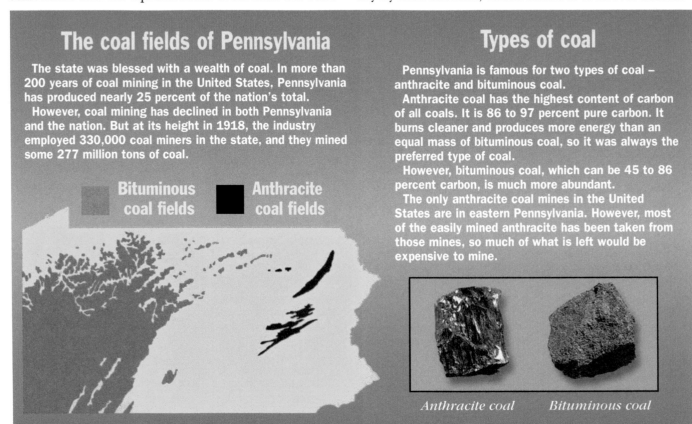

The coal fields of Pennsylvania

The state was blessed with a wealth of coal. In more than 200 years of coal mining in the United States, Pennsylvania has produced nearly 25 percent of the nation's total.

However, coal mining has declined in both Pennsylvania and the nation. But at its height in 1918, the industry employed 330,000 coal miners in the state, and they mined some 277 million tons of coal.

Bituminous coal fields   Anthracite coal fields

Types of coal

Pennsylvania is famous for two types of coal – anthracite and bituminous coal.

Anthracite coal has the highest content of carbon of all coals. It is 86 to 97 percent pure carbon. It burns cleaner and produces more energy than an equal mass of bituminous coal, so it was always the preferred type of coal.

However, bituminous coal, which can be 45 to 86 percent carbon, is much more abundant.

The only anthracite coal mines in the United States are in eastern Pennsylvania. However, most of the easily mined anthracite has been taken from those mines, so much of what is left would be expensive to mine.

Anthracite coal   *Bituminous coal*

of years, compressing the layers of plant matter beneath. Eventually, the layers of plant matter turned to coal.

Oil and natural gas also formed from dead living things. In this case, the living things are the small organisms that lived in the sea, such as bacteria, algae and plankton. Some 400 million years ago, western and much of central Pennsylvania were covered by a shallow inland sea and when the small sea creatures died, they collected at the bottom. Layer built up upon layer and eventually the layers were buried by sand and gravel that turned to rock through the centuries. Under great pressure and heat deeper in the earth, the layers of dead sea creatures turned to oil. At even higher temperatures, they turned to natural gas.

Some types of rock, such as sandstone, limestone and shale, are porous, which means they have many holes in them (like a sponge does) and the holes allow liquids and gases, such as water, oil and natural gas, into them. When oil and gas formed underground, they would build up in these layers of porous bedrock, as would water (although oil, gas and water don't mix, so they would create separate pools or reservoirs). Because of pressure deeper in the Earth, the oil and gas would try to

A retouched photograph of the Drake oil well, with Edwin Drake, foreground, at the right

rise. But if there is rock around the porous rock that has no holes, the oil and gas become trapped and vast reservoirs of oil or gas can sometimes build up there. That is what the people who drill oil and gas wells look for – underground reservoirs of these fuels.

One of the richest supplies of natural gas in this country is located in a deposit called the Marcellus formation that lies beneath much of Pennsylvania, Ohio and West Virginia. It's estimated there is enough gas contained in the shale that makes up the formation to supply all the natural gas needed in the United States for more than a decade.

However, most of the gas can only be recovered though a controversial process called "fracking" in which water, sand and chemicals are injected at high pressure into the shale, sometimes more than a mile underground, in order to release the gas. Although the process would increase this country's fuel independence, some worry that it could lead to pollution of some underground fresh water supplies and even create minor earthquakes by disturbing ancient faults in the underlying bedrock.

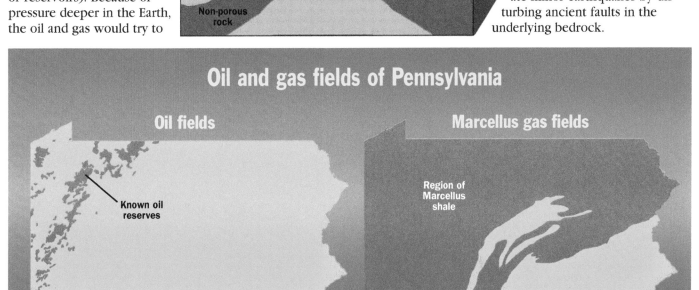

A pair of Diplodocuses watch a Pterodactyl soar overhead, a scene that likely depicts a shoreline in what is now western North America in the late Jurassic Period

Fearfully great lizards: translated from Greek, that's what "dinosauria" actually means. For nearly 165 million years, from about 230 million years ago to 65 million years ago, these legendary beasts roamed the Earth.

Wherever you walk in North America, the fierce tyrannosaurus rex and the massive triceratops once walked. But while scientists find the bones of these ancient creatures in other regions, they find few of them in Pennsylvania.

Bones are preserved as fossils when they are buried by sand or mud, which then turns to sedimentary rock. However, conditions in Pennsylvania were not ideal for preserving the sedimentary bedrock bearing these fossils. For most of the last 200 million years ago, the surface rocks of the state have been worn down by rain and wind, a process called erosion. However, geologists do find fossilized footprints of dinosaurs in Pennsylvania.

A rock bearing the footprint of a dinosaur is like a photograph that is millions of years old. It's a snapshot in stone of events of perhaps one warm summer morning by the shores of an ancient river or lake. A dinosaur might have wandered down to the edge of the water to drink, pressing its feet into the soft mud created by a rainstorm the evening before. In the hot sun, the mud and prints dried and hardened and were buried by other layers of mud containing still more dinosaur prints.

Over the centuries, the buried layers and prints would turn to rock such as shale. Erosion and earthquakes eventually brought some of the footprint-bearing rocks back to the surface, allowing a geologist to crack apart the layers to reveal the tracks, like someone turning the pages of a photo album.

From about 200 to 190 million years ago, conditions in what is today south-central Pennsylvania, in York and

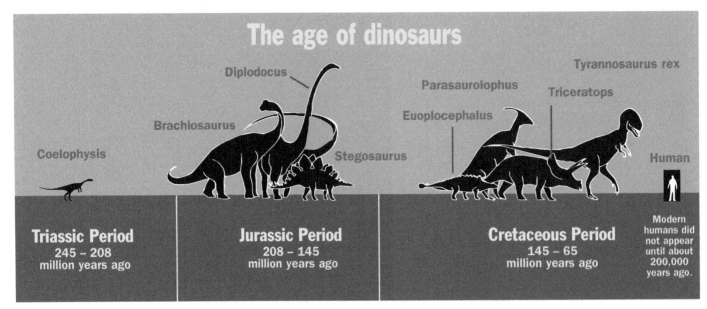

The age of dinosaurs

Diplodocus

Brachiosaurus

Coelophysis

Stegosaurus

Euoplocephalus

Parasaurolophus

Triceratops

Tyrannosaurus rex

Human

Triassic Period
245 – 208
million years ago

Jurassic Period
208 – 145
million years ago

Cretaceous Period
145 – 65
million years ago

Modern humans did not appear until about 200,000 years ago.

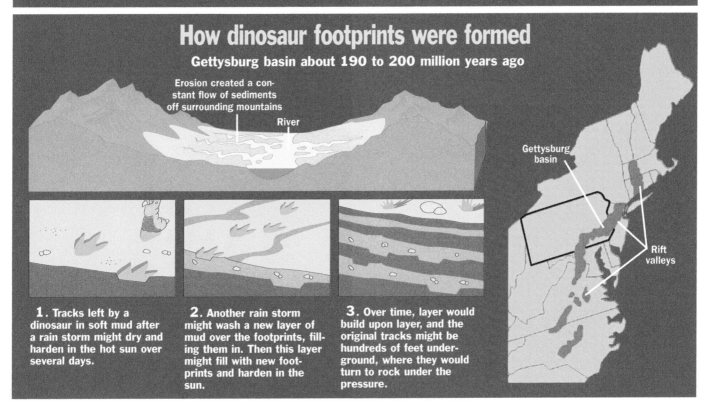

How dinosaur footprints were formed
Gettysburg basin about 190 to 200 million years ago

Erosion created a constant flow of sediments off surrounding mountains

River

Gettysburg basin

Rift valleys

1. Tracks left by a dinosaur in soft mud after a rain storm might dry and harden in the hot sun over several days.

2. Another rain storm might wash a new layer of mud over the footprints, filling them in. Then this layer might fill with new footprints and harden in the sun.

3. Over time, layer would build upon layer, and the original tracks might be hundreds of feet underground, where they would turn to rock under the pressure.

Adams counties, were just right for creating fossilized tracks.

As the African continent struggled to pull apart from North America, a series of valleys, called rift valleys, was formed in the United States from Massachusetts down to North Carolina as the land was ripped apart, a false start for the eventual separation that would come farther to the east and create the Atlantic Ocean. One such valley going through Pennsylvania is called by geologists the Gettysburg basin. It filled with mud, and dinosaurs left their footprints before the mud hardened.

However, that era was in the early part of the age of dinosaurs, before the larger and better known of their

Dinosaur print

breed, such as tyrannosaurus rex, had developed. So the footprints that have been found in the Gettysburg basin are from relatively small Triassic Period dinosaurs, such as the coelophysis, which was about the size of a large ostrich.

Dinosaurs became extinct about 65 million years ago after a massive meteor struck the Earth. It's believed that so much dust was blasted into the atmosphere, where it lingered for years, that the climate cooled dramatically, killing off much of the Earth's vegetation. Larger animals, such as dinosaurs, that required great amounts of vegetation or that fed on animals that did, could not survive.

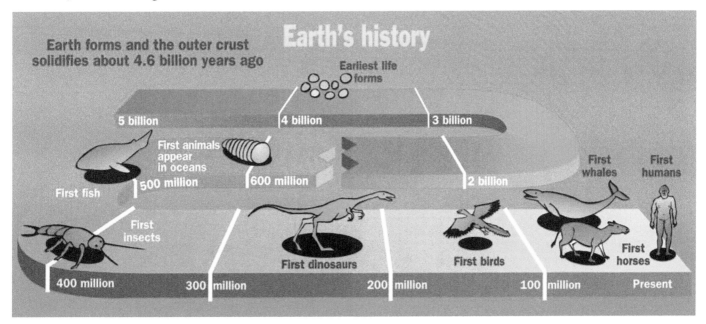

Earth's history

Earth forms and the outer crust solidifies about 4.6 billion years ago

Earliest life forms

5 billion

4 billion

3 billion

First animals appear in oceans

500 million

600 million

2 billion

First whales

First humans

First fish

First insects

First dinosaurs

First birds

First horses

400 million

300 million

200 million

100 million

Present

It was the dead of winter even in the heart of summer. It was the ice age.

During the last two million years, there have been four major advances of continental glaciers into the northern United States from centers in Canada.

The most recent ice age began about 70,000 years ago when a glacier that had formed in eastern Canada slowly expanded, eventually moving into the U.S. Northeast. The glacial ice, which was nearly a mile high in places, stopped advancing when it reached northern Pennsylvania, but that still created conditions in much of the rest of the state like a frozen arctic tundra.

The glacier, called by geologists the Laurentide ice sheet, also spread west, covering most of Canada and the northernmost United States from the Rocky Mountains eastward.

It's estimated the glacier was in portions of Pennsylvania for 5,000 years and that it did not begin to melt away until 21,000 to 22,000 years ago.

Glaciers form when the climate of a region cools enough so that snow builds up in winter faster than it melts in summer. Just as a snowball turns to an ice ball if you squeeze it hard enough, snow, if it continues to pile up, can eventually become so compressed that much of it turns to ice, especially at the bottom of the pile.

The increase in snow and ice year after year can create a mound of ice so high that the weight of it causes the ice to begin to spread slowly outward at the bottom, like thick maple syrup flowing on a tabletop – except at a much slower pace. On average, a glacier may advance a few feet a day. But at times, it may barely move and at other times it may surge forward several hundred feet in a day.

Even at this relatively slow speed, though, a glacier can move a great distance over thousands of years, as it did during the last ice age.

Much of the landscape of the Northeast was shaped by glaciers. If there had been no glaciers, the Great Lakes would not appear as they do today. And there would be no Cape Cod or Long Island.

A glacier can be like a bulldozer, widening and deepening valleys as it moves. Like a bulldozer, it may also push rocks and boulders along in front of it. When the ice melts, the meltwater may fill the valleys it dug, forming lakes. And the debris that was pushed along at the front

A family of woolly mammoths during the ice age.

edge of the glacier may be left in a long pile called a moraine.

The beds of the Great Lakes were ancient river valleys that were dramatically deepened by the action of glaciers. Cape Cod and Long Island were formed by glacial moraines. They sit above the ocean because of all the debris, including sand and pebbles, that the glacier dropped as it melted.

Pennsylvania, like much of the Northeast, shows the evidence of the glaciers' presence those thousands of years ago. In the areas that were covered by glaciers, the soil is filled with rocks and boulders that had been carried by the ice then dropped as it melted. Farmers in these areas know this all too well. Where the land was not covered by glaciers, the soil is largely free of the same kind of debris, which tells scientists where the glacier stopped in its advance.

However, once the ice was gone from Pennsylvania, the landscape did not immediately spring back to life. It was probably frozen, barren and blanketed with snow much of the year. Low tundra vegetation, similar to what you would see near the Arctic today, grew on it for perhaps centuries.

The first human beings arriving in the Northeast after the last ice age may have encountered very cold conditions indeed.

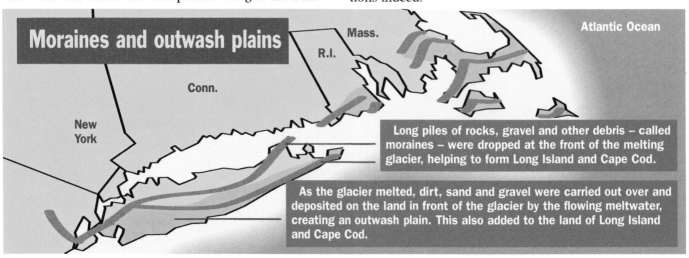

Moraines and outwash plains

Atlantic Ocean

Mass.

R.I.

Conn.

New York

Long piles of rocks, gravel and other debris – called moraines – were dropped at the front of the melting glacier, helping to form Long Island and Cape Cod.

As the glacier melted, dirt, sand and gravel were carried out over and deposited on the land in front of the glacier by the flowing meltwater, creating an outwash plain. This also added to the land of Long Island and Cape Cod.

North America's last glacier
(about 20,000 years ago)

Cordilleran
ice sheet

Laurentide
ice sheet

THE
FUTURE
CANADA

Glacier

Pennsylvania

THE FUTURE
UNITED STATES

Animals of the ice age

Giant cats with teeth like daggers. Shaggy-coated, long-tusked creatures that looked like elephants. Beavers larger than a man. Bison weighing a ton and a half. Musk oxen, ground sloths, wild pigs and caribou.

Some of the animals that roamed North America as the last ice age ended may have seemed to belong more to the age of dinosaurs. The much colder climate of the northern parts of the continent 20,000 years ago provided a home to a different group of animals than one sees today, creatures that were able to survive on a frozen landscape.

However, as the ice melted away and the climate warmed, many of the same animals found in the

Northeast today, including skunks, white-tailed deer, opossums, raccoons and black bears, moved into the region from warmer areas to the south.

It's believed the first humans arrived in North America between 13,000 and 14,000 years ago as the last ice age was ending. They may have hunted the largest of the ancient ice age animals to extinction in a matter of just 2,000 or 3,000 years, including the woolly mammoths and mastodons.

Saber-toothed cats are believed to have lived throughout North and South America. These huge predators, which also hunted mammoths and mastodons, probably went extinct soon after the ice age ended, when their primary prey disappeared.

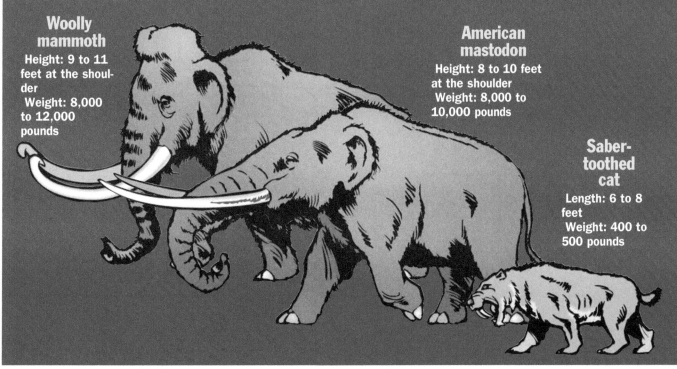

Woolly mammoth
Height: 9 to 11 feet at the shoulder
Weight: 8,000 to 12,000 pounds

American mastodon
Height: 8 to 10 feet at the shoulder
Weight: 8,000 to 10,000 pounds

Saber-toothed cat
Length: 6 to 8 feet
Weight: 400 to 500 pounds

The first Americans may have been a small band of hunters traveling in search of caribou or woolly mammoths more than 13,000 years ago who had no idea of the history they were about to make.

During the last ice age, so much of the Earth's water was ice that the surface of the ocean may have been more than 300 feet lower than it is today. As a result, a temporary land bridge emerged connecting Asia to Alaska. Human beings had already spread through much of Europe, Africa and Asia, but North America may still have been uninhabited by people.

The earliest North Americans may have entered Alaska from 17,000 to 40,000 years ago, but found their way south blocked by massive glaciers. However, it's believed that about 14,000 to 16,000 years ago, the climate warmed enough that a passage opened in the ice sheets allowing them to travel

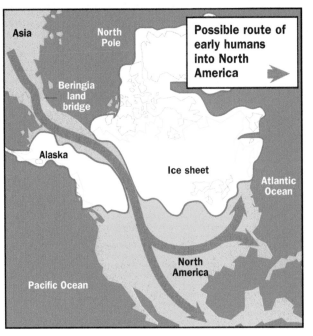

south into what is today the American Northwest. Others believe they may have bypassed the glaciers by traveling down the West Coast in boats much earlier.

Nevertheless, living for so long in frozen conditions, they may have been startled as they continued to travel south and east and began to encounter warmer temperatures and the vast forests and fields of America's heartland.

Archaeologists are detectives, piecing together clues about ancient people from the things they left behind. Occasionally, they find tools, artwork, trash or the remains of the buildings of the earliest residents of the Northeast. However, these artifacts have often been scattered, broken or changed through the centuries, making it difficult to determine what the earliest people in the region were like or how they lived.

In the 1920s, stone arrowheads were found near Clovis,

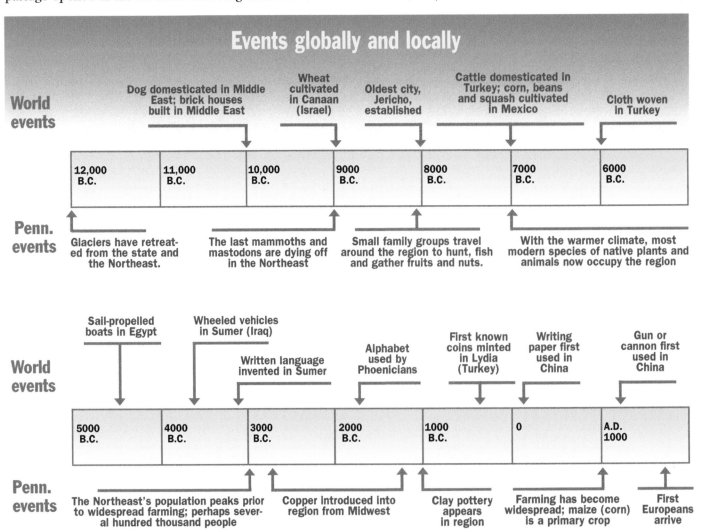

Events globally and locally

World events

Dog domesticated in Middle East; brick houses built in Middle East		Wheat cultivated in Canaan (Israel)	Oldest city, Jericho, established		Cattle domesticated in Turkey; corn, beans and squash cultivated in Mexico	Cloth woven in Turkey
12,000 B.C.	11,000 B.C.	10,000 B.C.	9000 B.C.	8000 B.C.	7000 B.C.	6000 B.C.

Penn. events

Glaciers have retreated from the state and the Northeast. — The last mammoths and mastodons are dying off in the Northeast — Small family groups travel around the region to hunt, fish and gather fruits and nuts. — With the warmer climate, most modern species of native plants and animals now occupy the region

World events

Sail-propelled boats in Egypt	Wheeled vehicles in Sumer (Iraq)	Written language invented in Sumer	Alphabet used by Phoenicians	First known coins minted in Lydia (Turkey)	Writing paper first used in China	Gun or cannon first used in China
5000 B.C.	4000 B.C.	3000 B.C.	2000 B.C.	1000 B.C.	0	A.D. 1000

Penn. events

The Northeast's population peaks prior to widespread farming; perhaps several hundred thousand people — Copper introduced into region from Midwest — Clay pottery appears in region — Farming has become widespread; maize (corn) is a primary crop — First Europeans arrive

New Mexico, that date back perhaps 13,000 years. For many years, the Clovis date was accepted as the earliest date for humans in North America. However, evidence was discovered in the 1970s for a human presence in Pennsylvania that may be earlier than the Clovis dates.

The Meadowcroft Rockshelter is a shallow cave overlooking a creek in Washington County in southwestern Pennsylvania. Radiocarbon dating, which is a method for measuring the age of carbon in things like charcoal and other once-living material, found that the site may have been used by humans 16,000 years ago and possibly as early as 19,000 years ago. However, not all archaeologists accept these earlier dates, believing the radiocarbon readings may be flawed.

Adding to the controversy is that sites that also have pre-Clovis dates have been found in other locations, including Virginia, Oregon, South Carolina and Texas. So the question of when humans first reached the Americas is not settled.

Nevertheless, the first arrivals in North America, whom we now call "paleoindians," which means the oldest Indians, probably had lives some people today would envy. After all, many may have had a wealth of food, land to live on and leisure time.

Were they savages? Not at all, say anthropologists. They were just as intelligent and inventive as people today. Human beings have probably been thinking at our modern level of complexity for the past 50,000 years. That means there was just as much genius among these early people as there is among modern humans. There was also just as much greed, charity, cruelty and compassion then as now.

It's true that life for humans has changed dramatically over the centuries as inventions and discoveries have accumulated. But humans' basic emotional and intellectual makeup has changed very little. If these early human cultures did not advance much beyond the use of simple tools, it's because there was little need for them to advance. By many standards, they were "rich" people. Their needs could be easily met. Indeed, they may have enjoyed life as much as many people today.

Paleoindians were a people frequently on the move, following the migration paths of the herd animals, such as caribou, or the waterfowl, such as ducks and geese, that they hunted. They also fished and trapped small game animals, such as beavers and rabbits, and they gathered and ate fruits, berries, nuts and some plants. In fact, plant matter may have made up much of their diet.

They probably carried their limited belongings on sleds or toboggans drawn by domesticated dogs or pulled by members of the group. Their

tools included hammers, chisels, axes and awls (hole makers) made from rock, as well as other implements, such as sewing needles, made from bones or the ivory of mammoth tusks.

The sewing needle dates back at least 40,000 years. So the first Americans probably wore tailored clothes of furs and skins. And they probably lived in tents made of skins that had been loosely sewn together.

They would camp, possibly for weeks or months at a time, near where animals were known to seek food or water, for instance by a river, near a forest opening or around ponds or marshes. They would eventually move on, but if the hunting, fishing and berry picking were good enough, they might return to the same spot the next season.

The earliest residents of the Northeast probably had rich spiritual and social lives. Studies of modern-day hunting and gathering societies, such as those discovered in tropical rain forests, show they can be masters of conversation. Without television and books, talk is their main source of entertainment and information.

They also relied on speech to pass along the knowledge accumulated over thousands of years – effective herbal cures, tips for hunting game, and designs of tents or layouts of camps.

It is likely they passed their leisure time doing craftwork, playing with wooden toys or games, singing and dancing, and just absorbing the beautiful scenery.

As the number of people grew in a region, it is also likely that many families came together at certain times of the year for festivals where goods were traded, stories swapped and marriages arranged.

Certainly, medicine was not as advanced as it is today. But the simplicity of that era offered its own protection. The level of stress for many paleoindians was probably very low. With so much food and land, war was not yet common, although personal arguments and hostilities have always existed.

While more babies died in infancy than die today, those who survived childhood could often live into their forties or fifties. Native Americans did not suffer many of the medical problems that are now widespread, such as obesity, diabetes and high blood pressure, which often result from inactive lives and diets high in fat.

Indeed, when the first Europeans arrived in America, they were struck by the great health Native Americans seemed to enjoy.

What was life like for the Native Americans of the Northeast? For the centuries before Europeans settled widely in the region, life was simple in many respects and it was cyclical. It had the seasons as its clock.

Many Native American families would live by the seashore or a riverbank in summer, to fish and

Fishing line

Hooks were made of bone, and lines were weighted with a rock to gather food in nearby fields and forests. In winter, those by the coast might move to protected inland areas to hunt squirrels, beavers, deer, moose or bears and to live off stored supplies of vegetables, nuts and berries.

Farming became common in the Northeast about 1,000 years ago. Crops included maize (corn), beans, squash, pumpkins, cucumbers and tobacco. Villages often sat beside the agricultural fields. And a protective fence might be built around the edge of the village so that it could be a winter home or a fort.

To create open spaces for farming and for foot paths, Native Americans of the Northeast might periodically burn over the land, destroying the underbrush and many of the trees. The practice would also create more open land for hunting, and it would kill off insect pests near villages.

Throughout the Northeast, Native American women were responsible for raising children, preparing meals and tending the crops and the home. Men were hunters, warriors and craftsmen, making the tools, utensils and weapons needed for survival.

Native American children, especially boys, were encouraged at a young age to be bold and self-reliant. To prove he had reached manhood, a young male might be led blindfolded into the woods in winter, armed with a bow and arrow, a knife and a hatchet, and be expected to survive on his own until spring.

The most common clothing for both men and women in warm weather was the breech clout, a belt and cloth that performed the same function as a pair of shorts does today. In colder weather, sewn layers of animal skins might be worn, often of raccoon or fox, with the fur side of the skin against the body.

Ax
About 7 in. long; made perhaps 7,000 years ago.

When the first Europeans arrived in the region in the early 1600s, there may have been several hundred thousand Native Americans living in the Northeast.

However, diseases contracted from the earliest European trappers and traders – smallpox, measles, influenza, typhoid fever and tuberculosis – killed many Native Americans. They had no history of these new illnesses and little or no immunity to them.

The first epidemic of the European illnesses, believed to include smallpox, probably began in the early 1600s. By 1700, up to 90 percent of the Native Americans in some areas of the Northeast had died from the diseases.

Because of disease and wars with Europeans, by 1800 the Native American way of life, which had defined how humans lived in the region for nearly 100 centuries, had disappeared from much of the Northeast.

Crafting points for tools and weapons

Native Americans hunted game animals with spears and arrows and cleaned them with knives. The points on these tools and weapons were often made from rock called chert or flint. When chert was tapped with a sharp rock or other hard tool, flat flakes would come loose, allowing the rock to be fashioned into the desired shape.

Paleoindians would chisel grooves, called flutes, into the sides of the points so that once the point was finished, it might be tightly inserted into a notch in a stick. It was then tied into place with animal tendons, leather strips or other stringy material.

Points were sometimes made from other types of rock, including obsidian and quartzite.

Spear point
Made 5,000 to 6,000 years ago

Arrowhead
Made 500 to 2,000 years ago

Knife point
Made 5,000 to 7,000 years ago

The tribes of Pennsylvania and New York about 1650

In upstate New York, five tribal groups banded together, perhaps as early as A.D. 1500, for mutual protection and greater collective strength. Known as the Iroquois Confederacy or League, the original five were the Mohawks, who lived in the Mohawk River Valley, and the Senecas, Cayugas, Onondagas and Oneidas, who lived in central and western New York.

The Confederacy eventually expanded to the south and west, reaching as far as lands that are part of present-day Ohio, Kentucky and Virginia.

It's believed that In 1656, the Confederacy defeated and absorbed the lands and surviving members of the Erie tribe of western Pennsylvania. And in 1677, the Susquehannock tribe in central Pennsylvania was defeated and the survivors were brought into the Confederacy. The Wenro tribe, in western New York, was brought into the Confederacy about the same time.

In 1722, the Tuscarora, originally a tribe based in North Carolina, was granted admission as the sixth nation of the Confederacy.

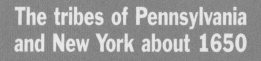

The Iroquois people often lived in farming villages made up of several longhouses surrounded by high wooden fences. Longhouses might be 50 to 200 feet long, 20 feet high and 20 feet wide. They were constructed from frameworks of wooden poles that were covered with bark.

The fences that surrounded the longhouses were made of long poles stuck in the ground. They served as protection in time of war, but they also kept out animals, such as wolves, and were a barrier against winter winds.

Inside a large longhouse, 15 to 20 related families might live, each with its own section of the open interior. Fires would be spaced throughout the structure, with smoke escaping through holes in the ceiling. Each fire might be shared by two families.

Longhouses could be smoky and noisy, with children and dogs scampering about. In summer, people would sleep or sit on raised wooden platforms. In winter, they might sleep on woven reed mats laid by the fires, perhaps wrapped in blankets of animal furs.

The Iroquois were farmers, fishers, hunters and gatherers. As farmers, they grew corn, beans and squash, called "the three sisters" and considered staples of their diet.

Food, such as corn or smoked fish, would be stored in large wooden tubs or casks. Valuable personal items might be placed in ceramic jars and then buried in the ground beneath sleeping areas. Other items might be placed on shelves above the sleeping platforms.

The Iroquois Confederacy was headed by a Grand Council made up of sachems who were elected by the women in each tribe. If they weren't doing a good job, sachems could be voted out of office.

After the arrival of Europeans in North America, disease, warfare, and squabbles and political disagreements among the nations of the Confederacy gradually took their toll. Traditional lands were lost and the Iroquois people became scattered, with many relocating to Canada.

However, the Iroquois have survived. It's estimated that more than 45,000 people who identify themselves as Iroquois live in Canada and nearly 80,000 live in the United States.

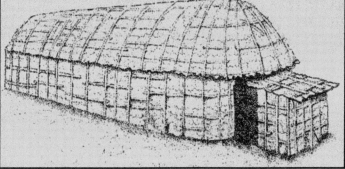

Iroquois longhouse

Some of the world's most magnificent mountains were once found in the Northeast. The Appalachian Mountains, which go through Pennsylvania, once rivaled the Alps or the Rocky Mountains in height.

But that was hundreds of millions of years ago. All those years of rain and wind, and the freeze and thaw of water in their crevices, have steadily eroded these peaks. Today, the region's mountains are minor when compared with those in other parts of the world.

Mount Davis, the highest peak in Pennsylvania, is 3,213 feet above sea level. Located in Somerset County in the southwest part of the state, it ranks 33rd on the list of the highest peaks in each state and it is less than one sixth the height of Mount McKinley in Alaska, the highest peak in the United States, at 20,327 feet above sea level.

View from the Appalachian Trail as it winds through Pennsylvania

The high peaks of Pennsylvania are concentrated in the southwest, in the Allegheny Mountains – part of the much larger Appalachian Mountain Range, which stretches from southeast Canada into central Alabama.

The northern tip of the Blue Ridge Mountains, which are also part of the Appalachian Mountain Range, is located in south-central Pennsylvania.

Generally, the taller the mountain is, the higher the erosion rate. For one thing, the climate at a higher altitude is usually harsher. But the key is not the elevation; it is the slope. And higher mountain peaks usually have steeper slopes. Water loaded with sediments will run down a steep slope faster, wearing away the rock beneath it at a faster rate. Erosion is by no means a steady process, though. A landslide may remove a huge chunk of a mountain in a day, lowering a peak by dozens of feet. But over a million years, even with the occasional landslide, the erosion rate may average only a fraction of an inch each year.

Many believe that Mount Everest, the highest peak in the world at 29,029 feet above sea level, has had a rate of erosion over the last million or so years of about a centimeter (.4 inches) a year. Interestingly, the rate at which the peak is falling from erosion may be about the rate at which it is still rising, leaving the overall height of Everest stable.

The erosion rate for Mount Davis is probably similar to that of other older mountain peaks in the Appalachians. One study in the Great Smoky Mountains estimated the rate there to be about .03 millimeters per year. That translates to erosion over a million years of about 100 feet.

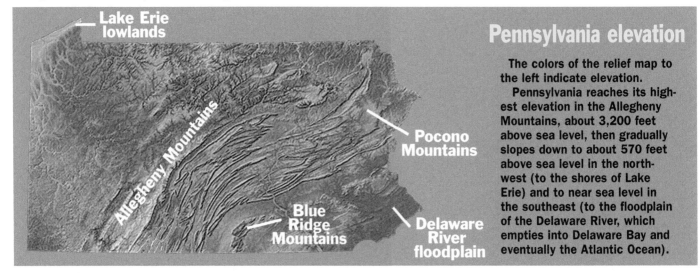

Lake Erie lowlands

Allegheny Mountains

Pocono Mountains

Blue Ridge Mountains

Delaware River floodplain

Pennsylvania elevation

The colors of the relief map to the left indicate elevation.

Pennsylvania reaches its highest elevation in the Allegheny Mountains, about 3,200 feet above sea level, then gradually slopes down to about 570 feet above sea level in the northwest (to the shores of Lake Erie) and to near sea level in the southeast (to the floodplain of the Delaware River, which empties into Delaware Bay and eventually the Atlantic Ocean).

The mountains of Pennsylvania

The ten highest peaks in Pennsylvania

(Height above sea level in feet)

1. Mount Davis – 3,213
2. Blue Knob – 3,120
3. Herman Point – 3,034
4. Schaefer Head – 2,950
5. Bald Knob – 2,930
6. Ritchey Knob – 2,865
7. Round Knob – 2,791
8. Round Top – 2,785
9. Packhorse Mountain – 2,766
10. Devies Mountain – 2,753

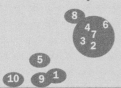

The Appalachian National Scenic Trail

Usually called just the Appalachian Trail, it extends from Springer Mountain in Georgia to Mount Katahdin in Maine, traveling about 2,200 miles through 14 states, including Pennsylvania. The first section opened in 1923 in New York. The trail is maintained by various trail clubs and private landowners along the way, and it is managed by the National Park Service and the Appalachian Trail Conservancy.

The highest point in each U.S. state

(Height above sea level in feet)

The highest point in North America is Mount McKinley in Alaska, 20,327 feet above sea level.

The highest point in the world is Mount Everest in Nepal, 29,029 feet above sea level.

The highest point in Pennsylvania is Mount Davis in Somerset County, 3,213 feet above sea level.

State	ft
AK: Mt. McKinley	20,327
CA: Mt. Whitney	14,494
CO: Mt. Elbert	14,433
WA: Mt. Rainier	14,410
WY: Gannett Pk.	13,804
HI: Mauna Kea	13,796
UT: Kings Pk.	13,528
NM: Wheeler Pk.	13,161
NV: Boundary Pk.	13,143
MT: Granite Pk.	12,799
ID: Borah Pk.	12,662
AZ: Humphreys Pk.	12,633
OR: Mt. Hood	11,239
TX: Guadalupe Pk.	8,749
SD: Harney Pk.	7,242
NC: Mt. Mitchell	6,684
TN: Clingmans Dome	6,643
NH: Mt. Washington	6,288
VA: Mt. Rogers	5,729
NE: Panorama Pt.	5,424
NY: Mt. Marcy	5,344
ME: Mt. Katahdin	5,267
OK: Black Mesa	4,973
WV: Spruce Knob	4,863
GA: Brasstown Bald	4,784
VT: Mt. Mansfield	4,393
KY: Black Mt.	4,145
KS: Mt. Sunflower	4,039
SC: Sassafras Mt.	3,560
ND: White Butte	3,506
MA: Mt. Greylock	3,491
MD: Backbone Mt.	3,360
PA: Mt. Davis	3,213
AR: Magazine Mt.	2,753
AL: Cheaha Mt.	2,407
CT: Mt. Frissell	2,380
MN: Eagle Mt.	2,301
MI: Mt. Arvon	1,979
WI: Timms Hill	1,951
NJ: High Point	1,803
MO: Taum Sauk Mt.	1,772
IA: Unnamed	1,670
OH: Campbell Hill	1,550
IN: Unnamed	1,257
IL: Charles Mound	1,235
RI: Jerimoth Hill	812
MS: Woodall Mt.	806
LA: Driskill Mt.	535
DE: Unnamed	442
FL: Unnamed	345

Water, water everywhere – and quite a lot to drink. Many people think of the Pacific Northwest as America's rain capital. In fact, Seattle gets less precipitation each year (an average of 38 inches) than Philadelphia (about 41.5 inches). Indeed, the average annual precipitation for Pennsylvania is 41 inches.

With more than 83,000 miles of rivers and streams and nearly 4,000 inland lakes and ponds, there is no shortage of water in the state. And that doesn't even include Lake Erie to the northwest. In all, 2.8 percent of the area of Pennsylvania is covered by water.

The presence of so much water produces a great diversity of wildlife. There is a long list of plants and animals that live only in water, and another long list of wildlife that live alongside water.

Pennsylvania gradually slopes from its higher central portion (Allegheny Mountains) to the northwest (the Lake Erie lowlands) and to the southeast (the Delaware River floodplain). So its rivers and streams generally follow those slopes.

Streams may empty into rivers which empty into larger rivers. The area containing all the rivers and streams that empty into one large river is called the watershed of the larger river. In the western part of the state, the final destination for that flowing water is often the Ohio River, which then flows to the Mississippi River and out into the Gulf of Mexico and the Atlantic Ocean. In central and eastern Pennsylvania, flowing water usually ends up in

The Schuylkill River as it passes by Philadelphia

the Susquehanna or Delaware rivers, and ultimately in the Atlantic Ocean.

Most Pennsylvania cities and towns get their water from reservoirs or from municipal wells that tap underground lakes of water, called aquifers. Aquifers are often found in underground layers of sand and gravel or in rock, such as limestone, that has holes, crevices or caves in it. Beneath the aquifer is solid rock through which the water cannot pass, so it acts like a bowl holding the water.

More than a billion gallons of water a day are drawn from aquifers across the state using wells, not only for drinking water but also for use by farms and businesses.

However, many rural residents of the state have private wells. With more than a million private wells, Pennsylvania is second only to Michigan in the number of such wells.

Lake or pond?

The usual definition of a pond is that it is shallow enough for aquatic plants to grow anywhere in it. But a lake can be so deep and dark in places that plants can only grow in the shallow areas.

Wetlands

Between dry land and the deep water of lakes, rivers or the ocean, you will often see wetlands, areas where land and water mix, such as swamps, marshes, bogs, wet meadows and riverbank forests that may flood after heavy rains.

Wetlands were once considered places of no value in much of America, and in the last two centuries more than half the acreage of wetlands in the lower 48 states has been filled, drained or otherwise lost. But wetlands do have great value, people have learned. They filter out pollution before it reaches larger bodies of water, and they are home to many kinds of plants and animals, from cattails to muskrats, that are seen almost nowhere else. That's why wetlands are now protected by law.

Since pre-Colonial times in Pennsylvania, about 56 percent of the state's wetlands have been lost, most of them inland freshwater wetlands. The state now has about 404,000 acres of freshwater and saltwater wetlands.

Watersheds

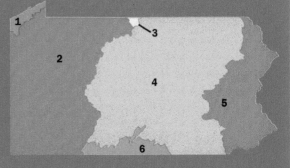

1 – Lake Erie
2 – Ohio River
3 – Genesee River
4 – Susquehanna River
5 – Delaware River
6 – Potomac River

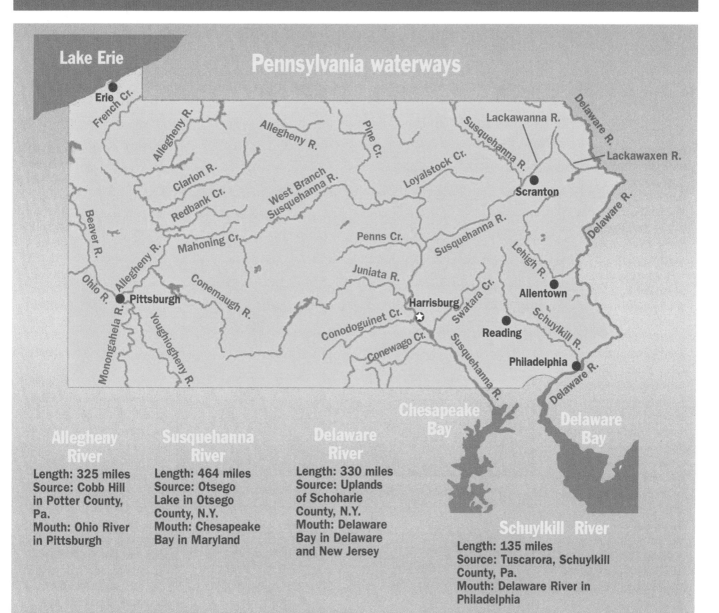

Pennsylvania waterways

Lake Erie

Allegheny River

Length: 325 miles
Source: Cobb Hill in Potter County, Pa.
Mouth: Ohio River in Pittsburgh

Susquehanna River

Length: 464 miles
Source: Otsego Lake in Otsego County, N.Y.
Mouth: Chesapeake Bay in Maryland

Delaware River

Length: 330 miles
Source: Uplands of Schoharie County, N.Y.
Mouth: Delaware Bay in Delaware and New Jersey

Schuylkill River

Length: 135 miles
Source: Tuscarora, Schuylkill County, Pa.
Mouth: Delaware River in Philadelphia

The Great Lakes

The Great Lakes, when combined, are the largest body of fresh water in the world, covering an area of 94,250 square miles, more than twice the area of Pennsylvania.

The Great Lakes were formed during recent ice ages when a series of advancing glaciers carved the lake basins. As the last glacier retreated about 12,000 years ago, the water from the melting ice filled the five massive lake beds.

Only Lake Michigan is entirely within U.S. borders. The other four lakes are shared with Canada. The lakes are connected together by rivers, straits and canals. The water flows from west to east, emptying into the St. Lawrence River and then into the Atlantic Ocean.

The Great Lakes system is so large that it may take more than two centuries for a drop of rain falling into Lake Superior to flow all the way to the ocean.

The smallest of the lakes by volume, Lake Erie is only 210 feet deep at its deepest point, and it may freeze over in winter. By contrast, Lake Superior, the largest of the lakes

and the largest freshwater lake in the world, reaches 1,333 feet in depth and normally does not freeze over.

An easy way to remember the names of the Great Lakes is to remember the word "homes" – Huron, Ontario, Michigan, Erie and Superior.

The Johnstown Flood

Johnstown
●

The rain that fell on Johnstown in late May of 1889 was unprecedented. By all accounts, six to 10 inches fell within a 36-hour period beginning on May 30, swelling creeks and sending rivers and streams that ran through the city over their banks. That forced many people to move their possessions and themselves to the upper floors of their homes, trapping them as the water in the streets continued to rise.

But the nightmare for those families was just beginning.

With a population of about 30,000, many of whom were immigrant steel workers, Johnstown was located in the lower elevations of the Conemaugh Valley. The city was also where the Little Conemaugh River and the Stonycreek River came together to form the Conemaugh River, which eventually flows into the Allegheny River.

In the hills above the city, about 14 miles away, sat Lake Conemaugh, a reservoir holding 20 million tons of water held back by the South Fork Dam. The lake and surrounding lands were part of the South Fork Fishing and Hunting Club, whose membership included many of Pittsburgh's wealthiest steel and coal industrialists and financiers, including Andrew Carnegie and Andrew Mellon. However, when the club bought the lake property in 1879, these rich individuals failed to improve the dam with water release pipes that could effectively lower

Johnstown's Main Street after the flood

The wreckage is searched for survivors and possessions

On May 31, after a day of torrential rain, the South Fork Dam collapsed at 3:10 p.m. The wave of water swept down through the valley, overwhelming small communities in its path – First South Fork, then Mineral Point, East Conemaugh and Woodvale. The wall of water picked up homes, barns, railroad cars and barbed wire, as well as animals and people. By the time the water reached Johnstown 57 minutes later, it was a surging tidal wave nearly 40 feet high and a half mile wide across the valley, and it pushed in front of it a mass of entangled debris that crushed whatever lay in its path.

One mother, trapped in her house with her seven children as the structure floated in the roiling flood waters, had to watch helplessly as all of them drowned. "What I suffered, with the bodies of my seven children floating around me in the gloom, can never be told," she later said.

Then, adding to the misery, as the water began to subside, the debris caught fire. By the time the water finally drained away from the city and the fire had run its course, Johnstown and the other communities between it and the failed dam were nearly unrecognizable. Some 2,209 people lay dead, including 99 entire families.

In the aftermath, news accounts of the destructive flood brought nearly $4 million in charitable donations from around the world (the equivalent today of more than $100 million). However, the wealthy members of the South Fork Fishing and Hunting Club were never held liable for the disaster

the water level behind the structure if it were to rise dangerously in a major storm. Also, they failed to maintain the earthen dam, sometimes patching cracks in it with only clay and straw. That would all prove fatal for so many people living in the valley below it.

as the court ruled it was an act of God that was directly responsible and not human negligence. American law would be changed by the seeming unfairness of the verdict, and today, in such circumstances, the membership would be made to pay.

Waterfalls

It's the contrast that's so striking – as if you've stumbled upon a three-ring circus in the middle of a desert.

Hiking through a forest, with the silence broken only by the occasional birdsong, one may gradually be aware of a vague roar of water ahead. Fifty yards more and it may turn eerily loud, totally out of place in such a quiet landscape. Waterfalls precede themselves that way.

Waterfalls are usually found where there is a considerable rise and fall to the land and where there is rain.

Waterfalls fall into categories. A cascade is a waterfall that typically descends gradually, in a series of small steps. It may be just a brook or stream moving down a slope of rocks. A cataract is a waterfall with a single, sheer drop that usually involves a large volume of water. What aren't cataracts or cascades fall back into the general category of waterfalls.

The highest waterfall in the state is Raymondskill Falls in the Delaware Water Gap National Recreation Area in Pike County. Its three tiers have a combined height of about 150 feet.

Raymondskill Falls in the Delaware Water Gap National Recreation Area

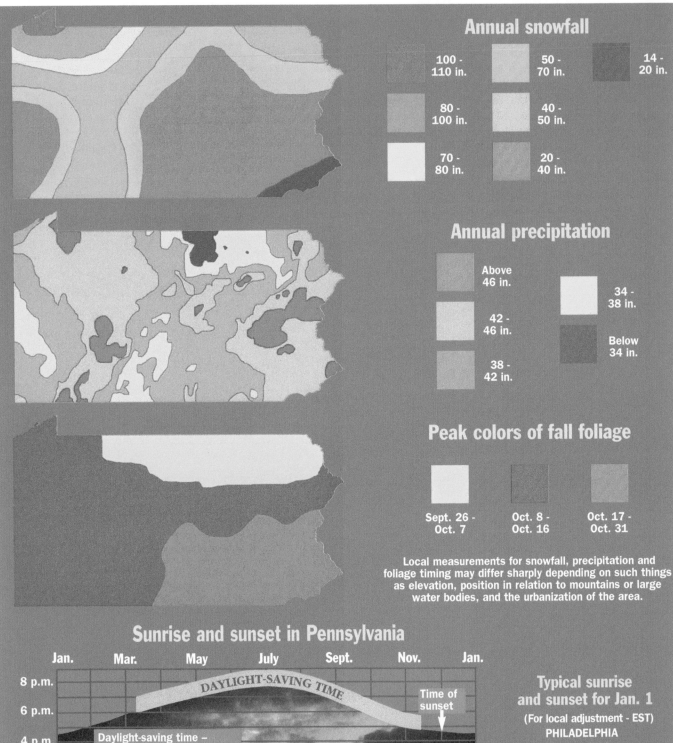

Annual snowfall

- 100 - 110 in.
- 50 - 70 in.
- 14 - 20 in.
- 80 - 100 in.
- 40 - 50 in.
- 70 - 80 in.
- 20 - 40 in.

Annual precipitation

- Above 46 in.
- 34 - 38 in.
- 42 - 46 in.
- Below 34 in.
- 38 - 42 in.

Peak colors of fall foliage

- Sept. 26 - Oct. 7
- Oct. 8 - Oct. 16
- Oct. 17 - Oct. 31

Local measurements for snowfall, precipitation and foliage timing may differ sharply depending on such things as elevation, position in relation to mountains or large water bodies, and the urbanization of the area.

Sunrise and sunset in Pennsylvania

Jan.   Mar.   May   July   Sept.   Nov.   Jan.

8 p.m.
6 p.m.
4 p.m.
2 p.m.
Noon
10 a.m.
8 a.m.
6 a.m.
4 a.m.

Feb.   Apr.   June   Aug.   Oct.   Dec.

DAYLIGHT-SAVING TIME

Time of sunset

Daylight-saving time – From the second Sunday in March to the first Sunday in November, the times for sunrise and sunset are set one hour later.

The fewest hours of daylight occur Dec. 21. Scranton – 9 hrs., 11 min.

The most hours of daylight occur June 21. Scranton – 15 hrs., 10 min.

Time of sunrise

DAYLIGHT-SAVING TIME

Typical sunrise and sunset for Jan. 1

(For local adjustment - EST)

PHILADELPHIA
7:22 a.m. 4:46 p.m.

READING
7:27 a.m. 4:48 p.m.

ALLENTOWN
7:26 a.m. 4:45 p.m.

SCRANTON
7:29 a.m. 4:44 p.m.

STATE COLLEGE
7:36 a.m. 4:54 p.m.

PITTSBURGH
7:43 a.m. 5:04 p.m.

ERIE
7:49 a.m. 4:59 p.m.

Philadelphia		Jan.	Feb.	Mar.	Apr.	May	June	July	Aug.	Sept.	Oct.	Nov.	Dec.	Total or annual avg.
RECORD TEMPS HIGH: 106 on Aug. 7, 1918 LOW: -11 on Feb. 9, 1934	Snowfall (inches)	6.5	8.8	2.9	0.5	0	0	0	0	0	0	0.3	3.4	22.4
	Precipitation (inches)	3.03	2.65	3.79	3.56	3.71	3.43	4.35	3.50	3.78	3.18	2.99	3.56	41.53
	Normal daily max. temp.	40.3	43.8	52.7	63.9	73.8	82.7	87.1	85.3	78.0	66.6	56.0	44.8	64.6
	Normal daily min. temp.	25.6	27.7	34.4	44.1	54.0	63.8	69.2	67.9	60.3	48.4	39.2	30.1	47.1
	Record high temp.	74	79	87	95	97	102	104	106	102	96	84	73	–
	Record low temp.	-7	-11	5	14	28	44	51	44	35	25	8	-5	–

Pittsburgh		Jan.	Feb.	Mar.	Apr.	May	June	July	Aug.	Sept.	Oct.	Nov.	Dec.	Total or annual avg.
RECORD TEMPS HIGH: 103 on July 15, 1988* LOW: -3 on Jan. 19, 1994*	Snowfall (inches)	11.5	10.2	7.4	1.5	0	0	0	0	0	0.4	2.1	8.3	41.4
	Precipitation (inches)	2.70	2.39	2.95	3.11	3.95	4.30	3.83	3.48	3.11	2.29	3.23	2.85	38.19
	Normal daily max. temp.	35.7	39.3	49.2	61.7	70.8	79.1	82.5	81.4	74.3	62.6	51.2	39.4	60.6
	Normal daily min. temp.	21.1	23.0	30.0	40.2	49.3	58.4	62.8	61.5	54.0	42.9	34.7	25.3	41.9
	Record high temp.	75	77	84	90	95	98	103	103	102	91	82	74	–
	Record low temp.	-22	-20	-5	11	26	34	42	39	31	16	-1	-12	–

Erie		Jan.	Feb.	Mar.	Apr.	May	June	July	Aug.	Sept.	Oct.	Nov.	Dec.	Total or annual avg.
RECORD TEMPS HIGH: 100 on June 25, 1988 LOW: -18, Jan. 19, 1994	Snowfall (inches)	29.8	18.1	13.8	3.1	0	0	0	0	0	.2	8.5	27.2	100.8
	Precipitation (inches)	2.94	2.39	2.95	3.33	3.44	3.76	3.54	3.46	4.61	4.05	3.93	3.83	42.21
	Normal daily max. temp.	33.7	35.5	43.8	56.1	66.6	75.7	79.8	78.6	71.9	60.8	49.9	38.1	57.5
	Normal daily min. temp.	20.8	21.1	27.5	38.1	48.2	58.4	63.5	62.5	55.8	45.3	36.6	26.6	42.0
	Record high temp.	72	75	82	89	90	100	99	94	99	88	81	75	–
	Record low temp.	-18	-18	-9	7	26	32	44	37	33	23	6	-11	–

Harrisburg		Jan.	Feb.	Mar.	Apr.	May	June	July	Aug.	Sept.	Oct.	Nov.	Dec.	Total or annual avg.
RECORD TEMPS HIGH: 107 on July 3, 1966 LOW: -22, Jan. 21, 1994	Snowfall (inches)	8.8	10.5	5.2	0.4	0	0	0	0	0	0	0.6	5.1	30.6
	Precipitation (inches)	2.88	2.39	3.37	3.10	3.79	3.60	4.61	3.20	4.07	3.27	3.23	3.23	40.74
	Normal daily max. temp.	37.0	40.7	50.4	62.4	72.1	81.0	85.5	83.4	75.6	64.1	53.1	41.3	62.2
	Normal daily min. temp.	22.8	25.1	33.0	41.9	52.1	62.0	66.3	64.5	56.2	44.6	35.1	26.6	44.2
	Record high temp.	73	78	87	93	97	100	107	104	102	97	84	75	–
	Record low temp.	-22	-13	-1	11	31	40	49	45	30	23	10	-8	–

* If ties, the date given is the most recent day on which the record was set   (Source: NOAA)

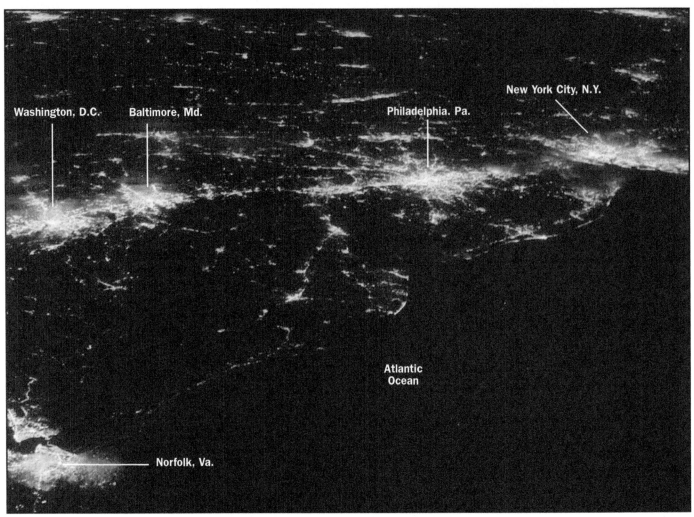

View of the East Coast at night from space (NASA)

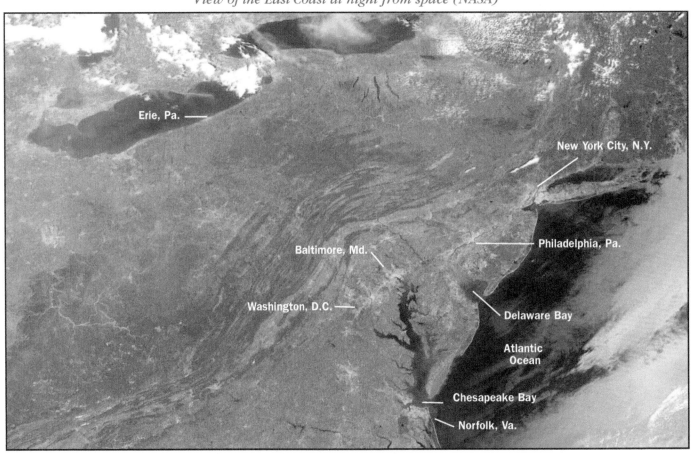

View of the Mid-Atlantic states in the day from space (NASA)

Winter in eastern York County

PENNSYLVANIA NATURE CALENDAR

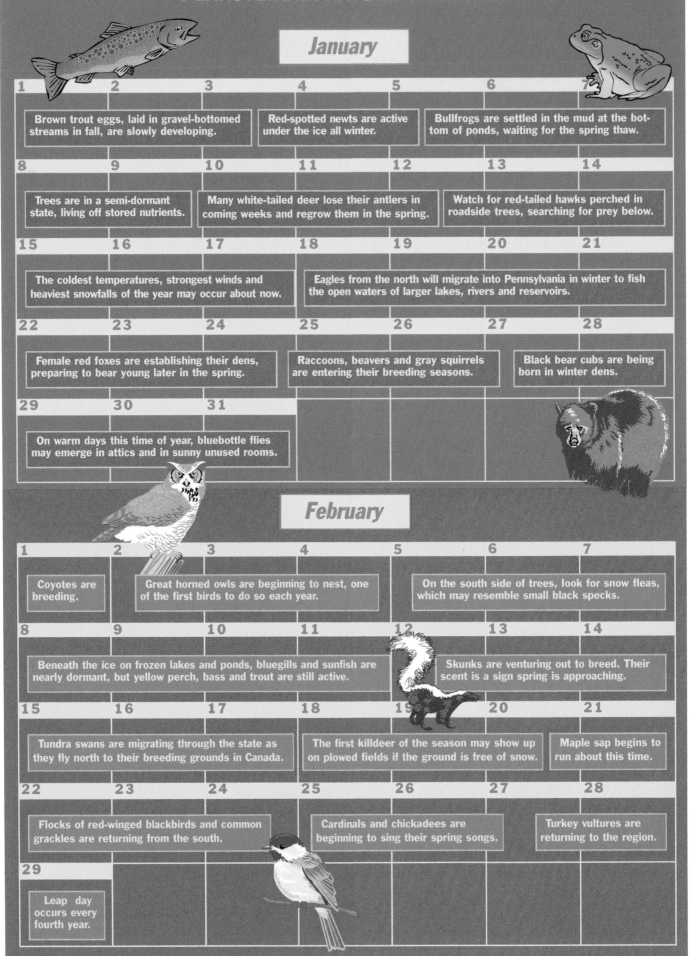

January

1	2	3	4	5	6	7
Brown trout eggs, laid in gravel-bottomed streams in fall, are slowly developing.			Red-spotted newts are active under the ice all winter.		Bullfrogs are settled in the mud at the bottom of ponds, waiting for the spring thaw.	

8	9	10	11	12	13	14
Trees are in a semi-dormant state, living off stored nutrients.			Many white-tailed deer lose their antlers in coming weeks and regrow them in the spring.		Watch for red-tailed hawks perched in roadside trees, searching for prey below.	

15	16	17	18	19	20	21
The coldest temperatures, strongest winds and heaviest snowfalls of the year may occur about now.			Eagles from the north will migrate into Pennsylvania in winter to fish the open waters of larger lakes, rivers and reservoirs.			

22	23	24	25	26	27	28
Female red foxes are establishing their dens, preparing to bear young later in the spring.			Raccoons, beavers and gray squirrels are entering their breeding seasons.		Black bear cubs are being born in winter dens.	

29	30	31
On warm days this time of year, bluebottle flies may emerge in attics and in sunny unused rooms.		

February

1	2	3	4	5	6	7
Coyotes are breeding.	Great horned owls are beginning to nest, one of the first birds to do so each year.			On the south side of trees, look for snow fleas, which may resemble small black specks.		

8	9	10	11	12	13	14
Beneath the ice on frozen lakes and ponds, bluegills and sunfish are nearly dormant, but yellow perch, bass and trout are still active.				Skunks are venturing out to breed. Their scent is a sign spring is approaching.		

15	16	17	18	19	20	21
Tundra swans are migrating through the state as they fly north to their breeding grounds in Canada.			The first killdeer of the season may show up on plowed fields if the ground is free of snow.		Maple sap begins to run about this time.	

22	23	24	25	26	27	28
Flocks of red-winged blackbirds and common grackles are returning from the south.			Cardinals and chickadees are beginning to sing their spring songs.		Turkey vultures are returning to the region.	

29
Leap day occurs every fourth year.

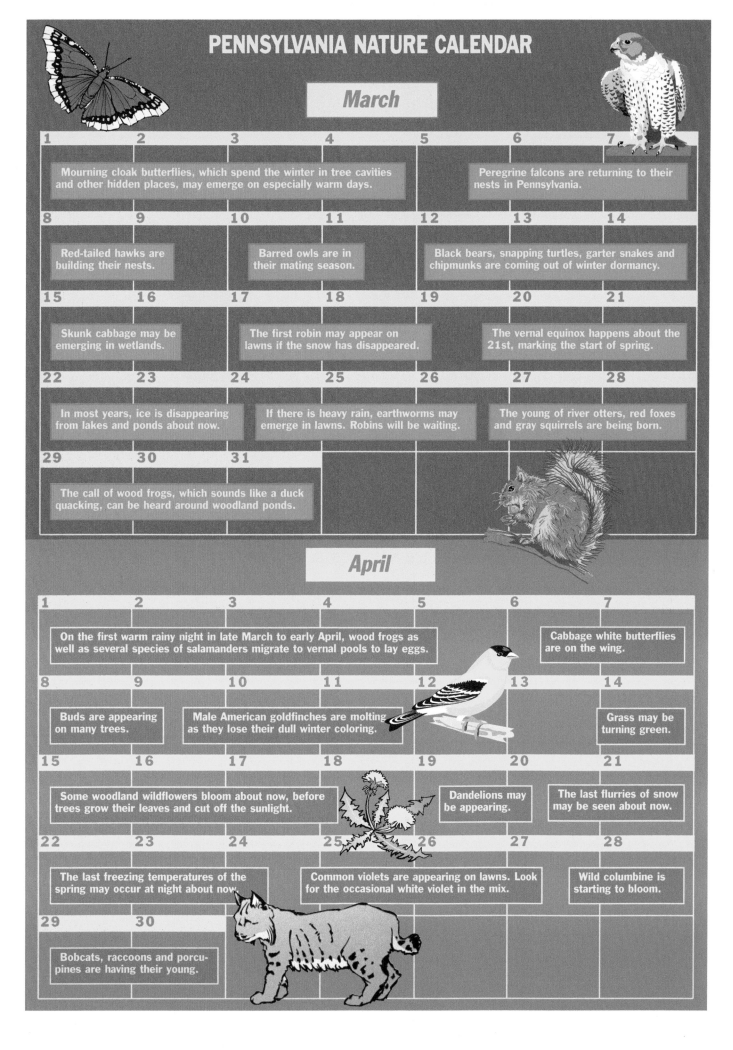

PENNSYLVANIA NATURE CALENDAR

March

| 1 | 2 | 3 | 4 | 5 | 6 | 7 |

Mourning cloak butterflies, which spend the winter in tree cavities and other hidden places, may emerge on especially warm days.

Peregrine falcons are returning to their nests in Pennsylvania.

| 8 | 9 | 10 | 11 | 12 | 13 | 14 |

Red-tailed hawks are building their nests.

Barred owls are in their mating season.

Black bears, snapping turtles, garter snakes and chipmunks are coming out of winter dormancy.

| 15 | 16 | 17 | 18 | 19 | 20 | 21 |

Skunk cabbage may be emerging in wetlands.

The first robin may appear on lawns if the snow has disappeared.

The vernal equinox happens about the 21st, marking the start of spring.

| 22 | 23 | 24 | 25 | 26 | 27 | 28 |

In most years, ice is disappearing from lakes and ponds about now.

If there is heavy rain, earthworms may emerge in lawns. Robins will be waiting.

The young of river otters, red foxes and gray squirrels are being born.

| 29 | 30 | 31 |

The call of wood frogs, which sounds like a duck quacking, can be heard around woodland ponds.

April

| 1 | 2 | 3 | 4 | 5 | 6 | 7 |

On the first warm rainy night in late March to early April, wood frogs as well as several species of salamanders migrate to vernal pools to lay eggs.

Cabbage white butterflies are on the wing.

| 8 | 9 | 10 | 11 | 12 | 13 | 14 |

Buds are appearing on many trees.

Male American goldfinches are molting as they lose their dull winter coloring.

Grass may be turning green.

| 15 | 16 | 17 | 18 | 19 | 20 | 21 |

Some woodland wildflowers bloom about now, before trees grow their leaves and cut off the sunlight.

Dandelions may be appearing.

The last flurries of snow may be seen about now.

| 22 | 23 | 24 | 25 | 26 | 27 | 28 |

The last freezing temperatures of the spring may occur at night about now.

Common violets are appearing on lawns. Look for the occasional white violet in the mix.

Wild columbine is starting to bloom.

| 29 | 30 |

Bobcats, raccoons and porcupines are having their young.

PENNSYLVANIA NATURE CALENDAR

May

1	2	3	4	5	6	7
Ruby-throated hummingbirds are returning from the south.		Snakes and turtles spend the day basking in the sun to warm their bodies.			The leaves on most trees have fully opened. Sugar maples will leaf out before red maples.	

8	9	10	11	12	13	14
Warblers have been arriving from the south in recent weeks.		Many songbirds, including robins, cardinals and chickadees, are laying eggs.			The young of mallards and other ducks are hatching. Within a day they will take to water.	

15	16	17	18	19	20	21
Jack-in-the-pulpits are blooming.			Many butterflies are out, including eastern black swallowtails and little wood satyrs.		Downy woodpeckers are laying their eggs in tree cavities.	

22	23	24	25	26	27	28
Wild lupines are flowering in meadows.		Wood turtles are laying eggs in areas of sand or gravel.				Violets are disappearing from lawns and fields.

29	30	31				
Treefrogs are in their breeding season and are most likely to be seen this time of year.						

June

1	2	3	4	5	6	7
Monarch butterflies are returning to the region.			White-tailed deer are having their young.		June bugs are emerging, often swarming around porch lights.	

8	9	10	11	12	13	14
Painted and snapping turtles are climbing out of ponds to lay their eggs.				Fragrant water lilies are blooming in ponds and Queen Anne's lace is flowering along roadsides.		

15	16	17	18	19	20	21
Black-eyed Susans, oxeye daisies and other field wildflowers are coming into bloom.				Day lilies are blooming.	The summer solstice happens about the 21st, marking the start of summer.	

22	23	24	25	26	27	28
Snakes that lay eggs, such as milk snakes, are doing so about now. Other snakes, like garter snakes, have live young later in the season.					The flashing of fireflies can be seen at night above fields and meadows.	

29	30					
Bears are in the heart of their breeding season.						

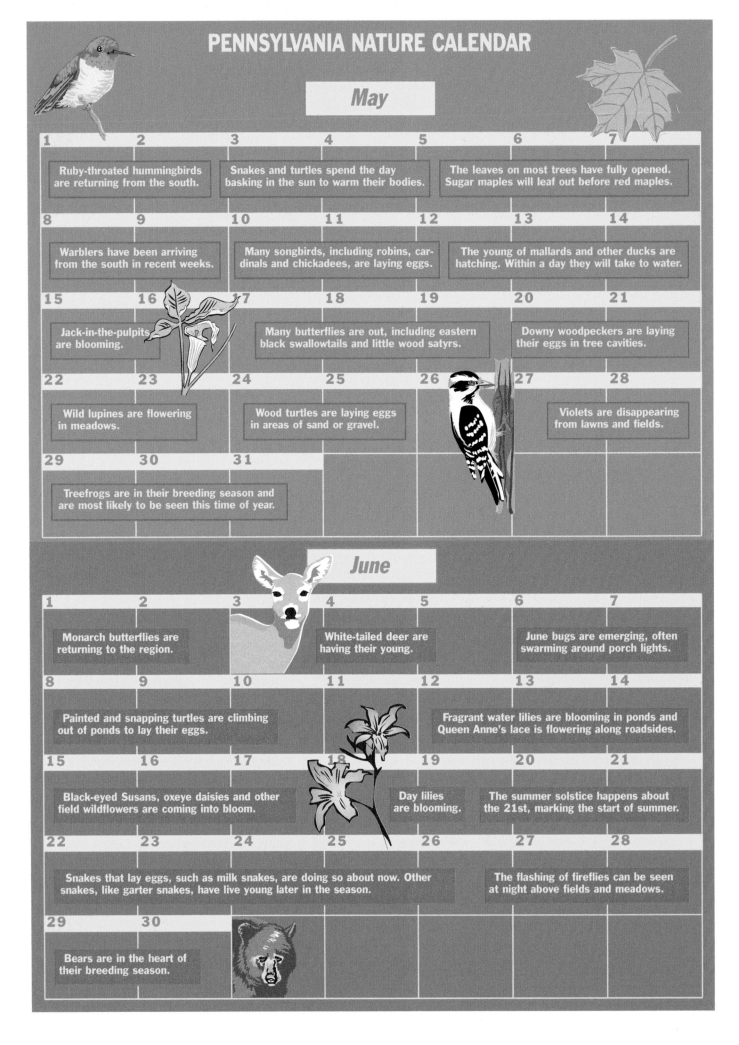

PENNSYLVANIA NATURE CALENDAR

July

1	2	3	4	5	6	7
The young of many raptors, such as red-tailed hawks and great horned owls, are fully grown and out on their own by now.				The season's second generation of clouded sulphur and spring azure butterflies is emerging.		

8	9	10	11	12	13	14
Butter and eggs is in bloom along roadsides and in meadows.		Wild blueberries are ripening.		The greatest amount of sunshine and the least cloudiness during daylight hours occurs about this time of year.		

15	16	17	18	19	20	21
Katydids and crickets can be heard calling at night.		The whine of cicadas can be heard during the heat of the day.				Tadpoles are emerging from green frog eggs.

22	23	24	25	26	27	28
The flutelike trill of gray treefrogs can be heard on cloudy days.		Chipmunks may be having their second litter of young this year.		The delicate orange flowers of jewelweed, also called touch-me-not, are in bloom.		

29	30	31
The hottest days of summer may occur about now and into early August.		

August

1	2	3	4	5	6	7
Goldfinches are beginning to nest, among the latest of any of the songbirds to do so.			With the breeding season for frogs over, ponds have become quiet.		Garter snakes may be having their young.	

8	9	10	11	12	13	14
Grasshopper and yellow jacket populations are increasing.		Shooting stars from the annual Perseid meteor shower can be seen at night.			Great blue herons have abandoned their nests for the season.	

15	16	17	18	19	20	21
Many types of goldenrod are blooming in fields and along roadsides. Joe-Pye weed is blooming around wetlands.				On rainy days, red-spotted newts can be seen on dirt roads in rural areas.		

22	23	24	25	26	27	28
The season's third generation of cabbage white butterflies is emerging.			The final heat wave of the summer may occur about now.		Leopard frogs can sometimes be seen in back yards.	

29	30	31
Tree swallows have moved to coastal areas, preparing to migrate.		

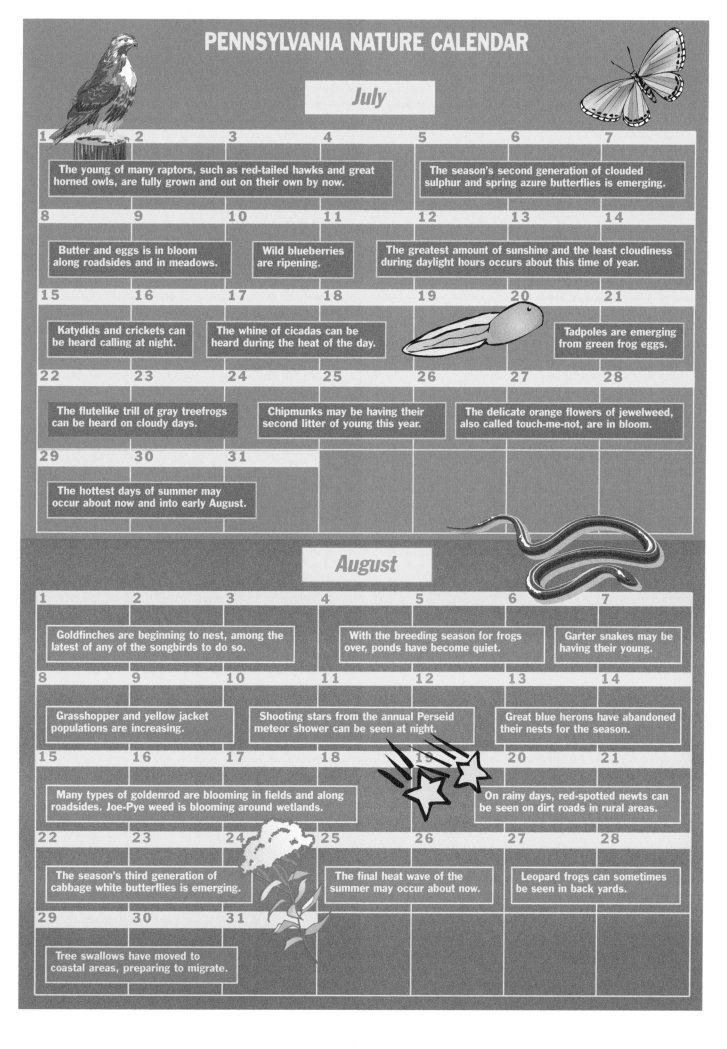

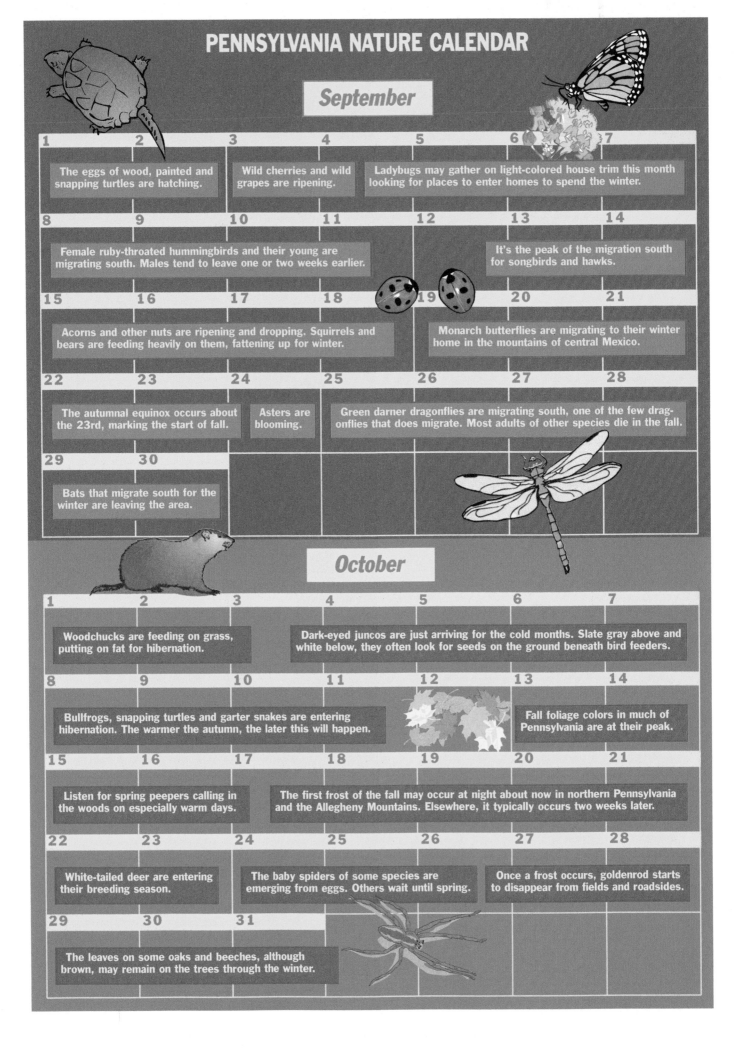

PENNSYLVANIA NATURE CALENDAR

September

1	2	3	4	5	6	7

The eggs of wood, painted and snapping turtles are hatching.

Wild cherries and wild grapes are ripening.

Ladybugs may gather on light-colored house trim this month looking for places to enter homes to spend the winter.

8	9	10	11	12	13	14

Female ruby-throated hummingbirds and their young are migrating south. Males tend to leave one or two weeks earlier.

It's the peak of the migration south for songbirds and hawks.

15	16	17	18	19	20	21

Acorns and other nuts are ripening and dropping. Squirrels and bears are feeding heavily on them, fattening up for winter.

Monarch butterflies are migrating to their winter home in the mountains of central Mexico.

22	23	24	25	26	27	28

The autumnal equinox occurs about the 23rd, marking the start of fall.

Asters are blooming.

Green darner dragonflies are migrating south, one of the few dragonflies that does migrate. Most adults of other species die in the fall.

29	30					

Bats that migrate south for the winter are leaving the area.

October

1	2	3	4	5	6	7

Woodchucks are feeding on grass, putting on fat for hibernation.

Dark-eyed juncos are just arriving for the cold months. Slate gray above and white below, they often look for seeds on the ground beneath bird feeders.

8	9	10	11	12	13	14

Bullfrogs, snapping turtles and garter snakes are entering hibernation. The warmer the autumn, the later this will happen.

Fall foliage colors in much of Pennsylvania are at their peak.

15	16	17	18	19	20	21

Listen for spring peepers calling in the woods on especially warm days.

The first frost of the fall may occur at night about now in northern Pennsylvania and the Allegheny Mountains. Elsewhere, it typically occurs two weeks later.

22	23	24	25	26	27	28

White-tailed deer are entering their breeding season.

The baby spiders of some species are emerging from eggs. Others wait until spring.

Once a frost occurs, goldenrod starts to disappear from fields and roadsides.

29	30	31				

The leaves on some oaks and beeches, although brown, may remain on the trees through the winter.

PENNSYLVANIA NATURE CALENDAR

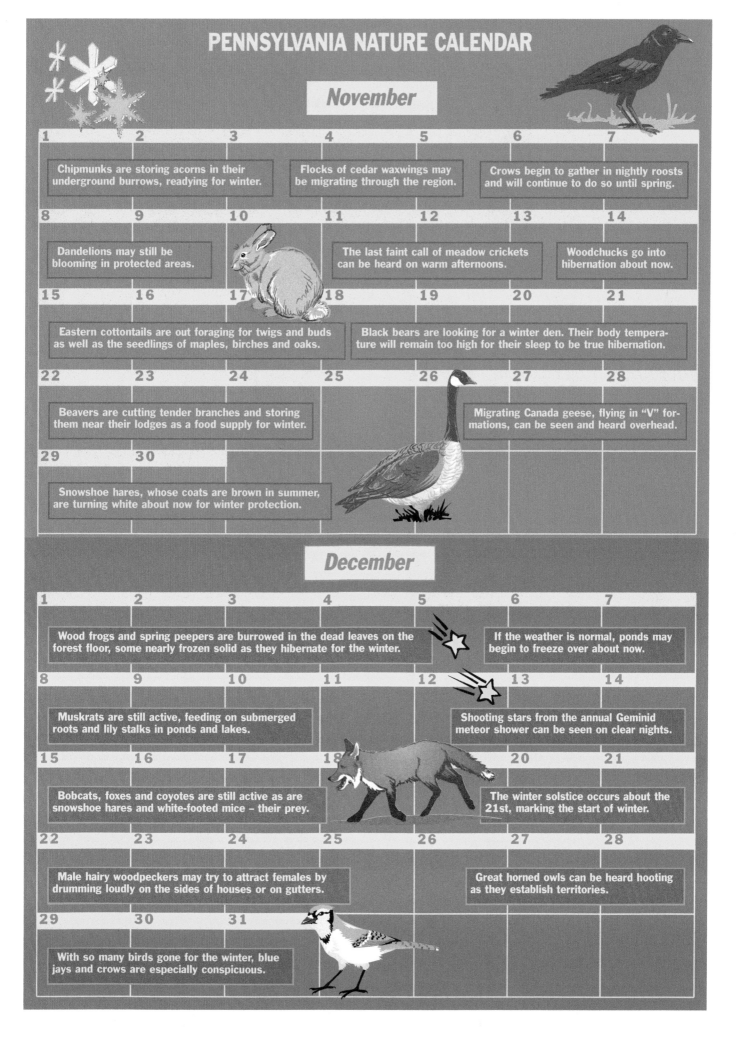

November

1	2	3	4	5	6	7

Chipmunks are storing acorns in their underground burrows, readying for winter.

Flocks of cedar waxwings may be migrating through the region.

Crows begin to gather in nightly roosts and will continue to do so until spring.

8	9	10	11	12	13	14

Dandelions may still be blooming in protected areas.

The last faint call of meadow crickets can be heard on warm afternoons.

Woodchucks go into hibernation about now.

15	16	17	18	19	20	21

Eastern cottontails are out foraging for twigs and buds as well as the seedlings of maples, birches and oaks.

Black bears are looking for a winter den. Their body temperature will remain too high for their sleep to be true hibernation.

22	23	24	25	26	27	28

Beavers are cutting tender branches and storing them near their lodges as a food supply for winter.

Migrating Canada geese, flying in "V" formations, can be seen and heard overhead.

29	30					

Snowshoe hares, whose coats are brown in summer, are turning white about now for winter protection.

December

1	2	3	4	5	6	7

Wood frogs and spring peepers are burrowed in the dead leaves on the forest floor, some nearly frozen solid as they hibernate for the winter.

If the weather is normal, ponds may begin to freeze over about now.

8	9	10	11	12	13	14

Muskrats are still active, feeding on submerged roots and lily stalks in ponds and lakes.

Shooting stars from the annual Geminid meteor shower can be seen on clear nights.

15	16	17	18	19	20	21

Bobcats, foxes and coyotes are still active as are snowshoe hares and white-footed mice – their prey.

The winter solstice occurs about the 21st, marking the start of winter.

22	23	24	25	26	27	28

Male hairy woodpeckers may try to attract females by drumming loudly on the sides of houses or on gutters.

Great horned owls can be heard hooting as they establish territories.

29	30	31				

With so many birds gone for the winter, blue jays and crows are especially conspicuous.

Peregrine falcon

The Natural History of Pennsylvania

Published by:
Hampshire House Publishing Co.
8 Nonotuck St.
Florence, Mass. 01062
www.hampshirehousepub.com

Printed in the USA
Copyright © 2015 by Stan Freeman and Mike Nasuti
ISBN: 97809893333-3-7

Some of the material in this book is based on articles by the authors that originally appeared in The Republican of Springfield, Mass.

If you find what you believe is an error, please notify us by email at hamphouse@comcast.net. The books are printed as they are ordered, which means we can upload a corrected digital file to the printer overnight and the new version will be in print the following day.

Facts and figures change, and every year around Jan. 1, we will update the book, even adding new photos or articles.

Photography credits

All photos are by Stan Freeman unless otherwise noted. A photograph's position on a page is indicated as follows:

T = top, C = center, B = bottom, R = right, L = left and A = all photos on page or in the indicated region of page.

Frandy Johnson – 28L; **Holly Schlacter Lowe** – 12T, 12CL, 12C; **Karel Jakubec** 4 (trout); **Illinois Department of Natural Resources** – 25TR, 26BL, 26BR, 32B, 35TC; **U.S. Centers for Disease Control** – 41BR, 55CL, 55CR; **U.S. Fish & Wildlife Service** – 23BL, 25 (gray wolf), 25 (red wolf), 31B, 33CL, 33CR, 36BC, 40TR, 53B, 70TL, 79, 110, front cover, back cover (lily); **NASA** – 102A; **National Park Service** – back cover (elk) **U.S, Department of Agriculture** – 59CR; **Library of Congress** – 84TR, 85TC, 98A; **Corel** – 11TL, 13A, 14A, 15TR, 15B, 19L (hawk), 21A, 22A, 24B, 25 (western coyote), 27B, 28BR, 28CR, 29A, 34BL, 38BL, 38BR, 39CR, 64CL, 64CR, 65CL, 111C; **Shutterstock** – 1TC, 20, 28TR, 30, 37, 45CL, 46TC, 50TR, 52B, 53TL, 54TR, 55BR, 59TR, 59CL, 62, 67, 71TC, 72T, 77TL, 77R, 86T, 88TR, 94TR, 96TR, 99, 103

Coyote howling at the moon

This series highlights the natural history of individual states in the Northeast. We began with a basic book, essentially a collection of articles, and we are attempting to adapt it to each state in the series. So some material in this book, including text, illustrations and photographs, is repeated from book to book in the series.

CPSIA information can be obtained at www.ICGtesting.com
Printed in the USA
BVOW07s0457021015

420265BV00009B/12/P